Information Governance and Assurance

Information Governance and Assurance

Editor

Fernand Qunicy

scitus academics

Information Governance and Assurance
Edited by **Fernand Qunicy**

ISBN: 978-1-68117-556-0
Library of Congress Control Number: 2016942965

www.scitusacademics.com
Box No. 4766, 616 Corporate Way,
Suite 2, Valley Cottage,
NY 10989

Notice

Reasonable efforts have been made to publish reliable data and views articulated in the chapters are those of the individual contributors, and not necessarily those of the editors or publishers. Editors or publishers are not responsible for the accuracy of the information in the published chapters or consequences of their use. The publisher believes no responsibility for any damage or grievance to the persons or property arising out of the use of any materials, instructions, methods or thoughts in the book. The editors and the publisher have attempted to trace the copyright holders of all material reproduced in this publication and apologize to copyright holders if permission has not been obtained. If any copyright holder has not been acknowledged, please write to us so we may rectify.

Printed in United States of America on Acid Free Paper

Preface

Information is an important asset in terms of the effective management and organization of resources and services. In a community where information is often regarded as a top-level asset, everyone is concerned about the protection and management of information. Information governance may be a set of controls, processes, procedures, policies and multi-disciplinary structures implemented to manage information at the company level, supporting the immediate and future regulatory, risk, legal, operational and environmental requirements of an organization. Info governance should determine the balance between two organizational goals that are potentially divergent: to extract value from information and to reduce the potential risk of information. It provides a framework to ensure that private information is handled legally, appropriately, effectively and efficiently in order to provide the best possible care. Information assurance (IA) could be a risk protection and risk management process for the use, transmission and storage of data and data systems. Information assurance processes typically ensure the following functions for data and related information systems: Availability ensures that information is ready for use by and at the necessary level of performance by individuals who are allowed access to it.

This book discusses the legal, organizational and ethical aspects of the governance and security of information and its relevance to all aspects of information work. This book offers practitioners and students the opportunity to practice good information management in real-world situations. This book will be a valuable resource for LIS students taking courses in information management and information governance, information professionals with a consulting or information management role within an organization.

Table of Contents

Chapter 1	Strategies for Information Management in Education: Some International Experience 1
Chapter 2	Electronic Resource Management Systems 33
Chapter 3	Developing an Electronic Resource Management System: Verde from Ex Libris 43
Chapter 4	Information Handling in Collaborative Research 57
Chapter 5	Knowledge Creation and Transfer Amongst Post-Graduate Students 71
Chapter 6	Establishing a Research Information System as Part of an Integrated Approach to Information Management: Best Practice at the Karlsruhe Institute of Technology (KIT) 87
Chapter 7	Aligning Web Content and Organisational Strategy: Towards Increasing Funding 99
Chapter 8	Framework to Manage Information Security for Malaysian Academic Environment 121
Chapter 9	Comparative Case Study on Website Traffic Generated by Search Engine Optimisation and a Pay-Per-Click Campaign, Versus Marketing Expenditure 143

Chapter 10 Information and Communications Technology Adoption Amongst Township Micro and Small Business: The Case of Soweto 169

Chapter 11 Convergence of Functional Areas in Information Operations 201

Chapter 12 Factors Influencing E-Collaboration for Knowledge Development and Innovation 221

Chapter 13 Beyond The Hype: Big Data Concepts, Methods, and Analytics 245

Chapter 14 Fusing Website Usability and Search Engine Optimization 269

Chapter 15 Use of Social Media Platforms for Improving Academic Performance at Further Education and Training Colleges 291

Index 311

Chapter 1

Strategies for Information Management in Education: Some International Experience

Andy Bytheway[1], Isabella M. Venter[1]

[1]*Department of Computer Science, University of the Western Cape, South Africa*

ABSTRACT

Background: Recent analysis of the management of information and communications technologies in South African education suggests strongly that there is only limited strategic thinking that might guide policy-makers, school principals, teachers, learners and suppliers of educational technologies. It is clear that here in South Africa, as elsewhere, the actual practice of technology-mediated education is driven more by the available technologies than by actual learner needs, good management principles and the wider national imperative. There might be lessons to be learned from experience elsewhere.

Objectives: This article reports and analyses conversation with eight international educators in Europe, Canada, the United States, New Zealand and Australia. All are managing the impact of technology in different ways (reactive and pro-active), at different levels (pre-primary through to senior citizen), in different roles (teachers, administrators and senior managers) and in different contexts (schools and universities).

Method: Open-ended conversations with educators and educational administrators in developed countries were recorded, transcribed and analysed. The qualitative analysis of the content was done in the style of 'open coding' and 'selective coding' using a qualitative content analysis tool.

Results: Whilst technology is still seen to drive much thinking, it is found that that success is not derived from the technology, but from a full and proper understanding of the needs and aspirations of those who are directly involved in educational processes, and by means of a managerial focus that properly recognises the context within which an institution exists.

Conclusion: Whilst this result might be expected, the detailed analysis of the findings further reveals the need to manage investments in educational technologies at different levels and in different ways.

INTRODUCTION

Education is rapidly becoming digitised at all levels, including the production, dissemination and transfer of knowledge. Despite the clear need for educators and learners to share knowledge, it is interesting that the initial impetus for introducing information technology (IT) into education often comes from the technology itself rather than from educational need. This can be seen as 'technology push', as opposed to the 'educational pull' that arises from a proper articulation of the needs of different educational activities and subjects (Anderson *et al.* 2002; Baker-Eveleth *et al.* 2007; Barron *et al.*2003). The technology push-pull arguments have prevailed for decades in business and the public sector, but it is now widely recognised and agreed that information and communications technology (ICT) investments should be justified on the basis of the benefits that will be gained (Ward & Daniel 2005). It follows that the benefits of ICT investments must be effectively managed in education, as they are elsewhere; as Ward and Daniel argue: in education (as in any kind of enterprise) the 'pull' is more important than the 'push'; managing *information* promises more benefit than managing*information technology*.

But, what exactly is it that sits between the achievement of the benefits of good information management and the technology that underpins it? And what is meant by 'information management'?

These are questions that all managers must ask, even in education. It is therefore surprising to find a lack of specific research into the management of ICT in education, despite the fact that these are times of great technology-induced change (Botha 2009; Motala 2009; Bytheway 2013). Change is happening fundamentally at the level of social behaviour (Angell 2000; Shirky 2008), and in education and knowledge management in South Africa and elsewhere (Moloi 2007; Omona, Van der Weide & Lubega 2010). In reading educational research, the word 'management' is not often found: a review of more than 700 articles published over the last 20 years found very few that specifically addressed the management of information technology and information systems in education; some are old and many are from developed regions of the world (Michael 1998); some point clearly to practice in regular management in business in order to find answers (Uys 2007:239), others argue for the use of systems analysis methods (Hardman & Paucar-Caceres 2010:168) or they verge on speculation (Bhusry & Ranjan 2012:315). When management is mentioned, it is often to decry the problems created by 'management', as if managers were separate from the real world of educational practice.

Of course, the degree of 'push' and 'pull' and the nature of the needed benefits depend on the context. In different countries significantly different levels of achievement might be found. For example, in Singapore it is now more than 25 years since the introduction of a system through which parents could check, each day, what their children had learned at school, using their television set linked up to the local education management systems – and that was before the general availability of the world wide web (Maslin 1990). At the other end of the scale, education in South Africa, a developing country, is reported to be in a precarious condition (Jansen 2012). In spite of many experts' belief that the introduction of ICTs into education will solve problems, experience with educational ICT initiatives in South Africa is variable, according to local research undertaken at different levels (Brown 2010; Bytheway 2004, Bytheway *et al.* 2010; Davids 2009; Koch 2006; Madiba 2009; Republic of South Africa 2004). A special problem is the induction of disadvantaged students with only a rural background, where cultural issues can present special challenges and where a free association with technology is found to be helpful to those students (Rahimi 2010). The educational 'pull' in a vibrant and successful context such as Singapore will have little in common with the needs of impoverished rural regions of Africa, where there might be no electricity and no data communications services.

In much of the expert analysis that is available, there is little that specifically addresses the *management* of information at a strategic level, or even at a systems level. The focus is commonly on pedagogy, didactics, specific technologies or contextual and social issues. The question arises: do we need to widen our viewpoint, and examine the educational processes and information resources that can be improved by better *information management*, rather than the generalities of *context* or the specifics of *information technology*? Hence, the central problem addressed here is that *the benefits of technology investments in education are not realised in South Africa*.

The problems in South Africa, and the evident belief that information technology offers solutions to some of our problems but has not really delivered them, partly because of the dependency on teacher attitudes (Chigona *et al.* 2011), led to the idea of a review of international experience with and attitudes to the management of digitised information and of information technology in education, by means of conversations with educators and administrators working in international education. This article explores how information technology is typically managed in selected countries, by an analysis of eight international conversations with educators and administrators that set out to explore the extent to which information and information technology are managed strategically in education, with what benefits to stakeholders.

Firstly, the article reviews the strategic management of information technology from a general perspective, drawing on theories from business management; a generic framework for the analysis of strategic alignment is introduced and then used to organise an open coding of the qualitative data derived from the eight conversations. A further selective coding and stakeholder analysis then examines the benefits of ICTs in education according to the different stakeholder groups, the benefits that they seek and the extent to which strategic planning might help.

The question at the heart of this work is: *does strategic planning help to deliver the benefits of information management in education*? As will become evident, 'information management' is considered to include a range of management activity, from technology through to strategy.

Strategic management of ICTs

There is extensive literature on strategic planning; in progressive businesses it is routine to undertake periodic analysis of internal and external challenges and opportunities, especially those relating to ICTs. The challenges and opportunities arising from information technology are a feature of much of that literature, but all organisations still find it difficult to achieve the anticipated results. Even in developed countries, in the public sector, huge amounts of money have been wasted in fruitless information technology projects (The London Independent 2010).

Difficulty prevails

Delivering benefits from any ICT investment is difficult, and it is all too easy to sweep failure under the 'carpet' of blame-games, shifting circumstances and general obfuscation about what is really needed. This is unfortunate, especially when there is such a long history of research and thought leadership about strategic information systems management. For at least 30 years this has been one of the hottest topics in the business research domain and its development can be traced over time (Zachman 1987; Baets 1992; Henderson & Venkatraman 1993; Edwards & Peppard 1997; Bytheway 1998; Luftman 2000; Kearns & Lederer 2000; Ward & Peppard 2002; Venkatesh *et al.* 2003; Ward & Daniel 2005) Over time the scope of research has broadened, for example to deal with other contextual issues that constrain success, such as human factors and organisational politics, something that the typical IT manager seems to find very difficult (Knights & Murray 1994; Chatham & Patching 2000)

Strategic management of ICTs in education

Strategic management is sometimes crudely reduced to a question with three parts: *where are we now, where do we want to be* and *how shall we get there?* The first real question, of course, is *where do we want to be?* This is where we debate and decide our objectives, the risks that are involved, the allocation of resources to strategic initiatives and so on. This then all depends on what people (those with influence) think about an organisation's strategic objectives and whether they perceive benefits that will accrue to their advantage. These influential people are the *stakeholders*, and these are the people we need to understand.

The remainder of this article presents an established framework for information management that joins issues of technology with issues of strategy and uses it to analyse the content of eight conversations with educators and educational administrators that illustrate how role-players in different countries are dealing with the management of ICTs in education. This provides some evidence with which to address the three questions above.

A framework for information management

During the 1990s, business managers and academics strove to find answers to critical questions about managing information in organisations. An early, extensive, review of literature concerning information systems 'success' was well received, is widely cited and has since been updated (DeLone & McLean 1992, 2003). More recently (as noted) there has been increasing attention to the management of benefits rather than actual or perceived success, but huge problems are still evident even in developed, supposedly capable countries (Chatterji 2007; The London Independent 2010; Ward & Daniel 2005). In the face of all this, how can the complexities of managing information in organisations be addressed?

Frameworks emerge

In the history of these matters, Zachman (1987) stands as a seminal source, but his frequently cited framework (a six-by-six matrix) was too complex for many managers to work with, having six layers of technology management down one side and six different perspectives on the business across the top: 36 different points of concern (the intersection of the six rows and six columns of the matrix) were just too many for many busy managers to handle.

Others have sought simpler views of the problem, and another frequently cited example is the Henderson and Venkatraman (1993) framework. It relates business and IT issues at the internal and external levels using a two-by-two matrix. This model is attractively simple at first sight, but it leads to a range of unanswered questions and lacks the elegance and symmetry that makes these things memorable. It implies dependencies and relationships between its conceptual components, but these are not immediately evident on a first reading, and as these authors admit in the original presentation of the idea, one has to go elsewhere to get the detail that is needed to operationalise the framework.

The information management body of knowledge

When these and other models are analysed and combined in order to understand what we need to be able to do well, a pattern emerges:

- **Information technology:** which comprises the requisite infrastructure for systems
- **Information systems**: which support an organisation's operational processes.
- **Organisational processes:** which deliver benefits through improved performance.

- **Organisational benefits:** that are sought by stakeholders
- **Organisational strategy:** that is realised by the delivery of required benefits (Bytheway 2004).

This arrangement of ideas has been developed, refined and tested. It is familiarly known as the 'IMBOK' (Information management body of knowledge). Recently it has been used to assess the perceived information management competencies of a large South African bank, and there is an archive of data from other sources that stands as a resource for future research (Bytheway 2011).

However, here we will use it in its simplest form, to provide a structure for the analysis of the international conversations. Figure 1 provides a simple overview.

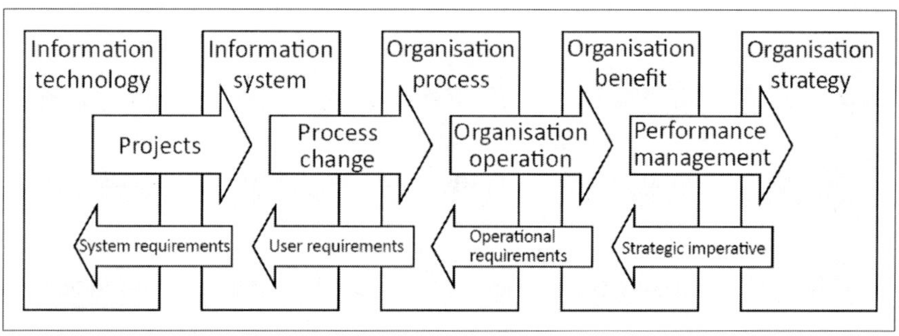

FIGURE 1: The information management body of knowledge (IMBOK).

As can be seen, the IMBOK framework divides the management domain into five (shaded) areas: technology, systems, processes, benefits and strategy. At the four intersections (the arrows) are the components of 'value' transfer and delivery: projects, process change, operations and performance; also, the imperatives: strategic imperatives, operational targets, user requirements and systems requirements.

So, to understand ways to maximise the benefits of investments in educational ICTs, we can now ask:

- What technologies are used in education?
- What systems assist education activities?

- What educational processes prevail and how are they changed by ICTs?
- What are the benefits and how are they measured?
- What strategies guide the use of ICTs in education?

Method of working

In a trip around the world, seven open-ended conversations with educators and educational administrators in developed countries were recorded (with permission) and have since been transcribed and analysed; one later conversation with an education assessment consultant in the United Kingdom makes up the eight (see Table 1).

TABLE 1: The eight conversations.

Person	Organisation	Location
Engineer	Senior citizens club	California, USA
Technology teacher	High school	New Zealand
Learning management system manager	Business faculty	New Zealand
Senior academic	University	Ohio, USA
Teacher	Primary school	Minnesota, USA
Teacher	Kindergarten	Australia
Learning chair	Business school	Canada
Consultant	Education consultancy	York, UK

The general approach adopted for the analysis of the content was in the style of 'open coding' and 'selective coding' (Strauss & Corbin 1998). Space precludes a detailed discussion of this qualitative approach to research, but the paragraphs below make it clear: they first present the open coding, by way of summary, and then the selective coding, based upon the domains of the IMBOK as described above.

Open coding

The first reading of the interview transcripts led to 26 a *posteriori* categories, as shown in Figure 2. On review, these categories were grouped into sets relating to:

- Activities of different kinds within education (principally administrative or educational).
- Contextual issues at the level of education, change and society at large
- The domains of the IMBOK.
- Issues open to further investigation.
- Resources of different kinds at different levels.
- Stakeholder references, indicating who is involved.

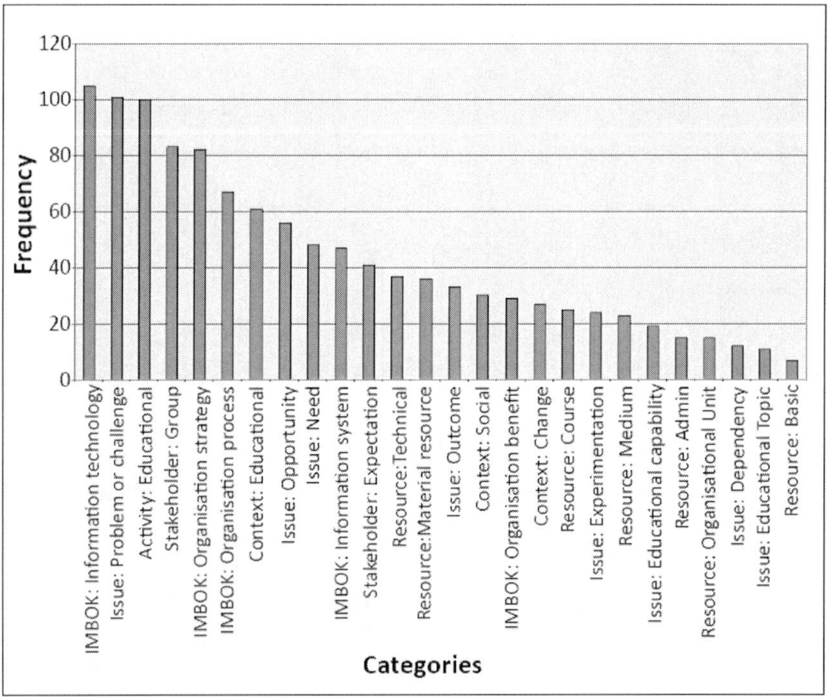

FIGURE 2: Open coding, 26 categories.

There was also a catch-all category of 'Others' that accommodated some of the academic references to theories of education and technology management.

The chart in Figure 2 tabulates the number of instances of coding of each category, across all of the interviews. Of course, to just count the frequency of occurrence of categories is a simple approach to analysis, but it provides a first view of the nature and content of the conversations and the focus of the respondents' interests.

As can be seen, a range of categories emerged from the open coding analysis, the most frequently occurring being*information technology* and the least frequent *basic resources*. Here we are principally concerned with those categories that map to the five layers of the IMBOK, as indicated by the 'IMBOK' prefix in the chart. Note therefore that the evidence presented here will be very selective because of the focus on the IMBOK as an analytical framework (and because of space constraints); the analysis does not represent the full richness of the 1139 actual instances of categorisation, and the 26 768 words that comprise the substance of the eight conversations.

Interesting findings from the open coding include the following observations:

- Information technology drove much of the conversation, overriding issues of strategy, process, systems and (most worryingly?) *benefits*.
- There were many more *problems* than benefits articulated by the respondents.
- As is often found in these kinds of conversation, discussion of technologies and problems were in the context of the educational *processes and activities* wherein they are found, rather than the deeper issues of benefits, experimentation and capability.
- *Opportunities*, *needs*, *expectations* and *resources* were all represented in the middle range of frequencies.

Hence, we find that there is a range of issues that extends from technology to strategic benefits, but the bias towards technology 'push' is strong.

The selective coding drives the analysis down into the issues of alignment, where we can see more clearly how information is managed, what benefits there are and whether there is any real strategic context within which management takes place.

Selective coding

From the 26 categories that emerged in the open coding, there are five relating directly to the five domains of the IMBOK:

- technology
- systems
- processes
- benefits
- strategies.

These five domains provide a useful framework with which to examine strategic management issues, and the principal portion of the analysis that follows is based on these five domains. It is useful to start by presenting the selected domains and their frequencies in the coding, this time in the same sequence in which the IMBOK presents them (see Figure 3). This allows us to see the extent to which the conversations dwelt upon these five principal domains. We can see these as five*specific* discourses within the general discourse about education and the way that information is managed therein. In summary:

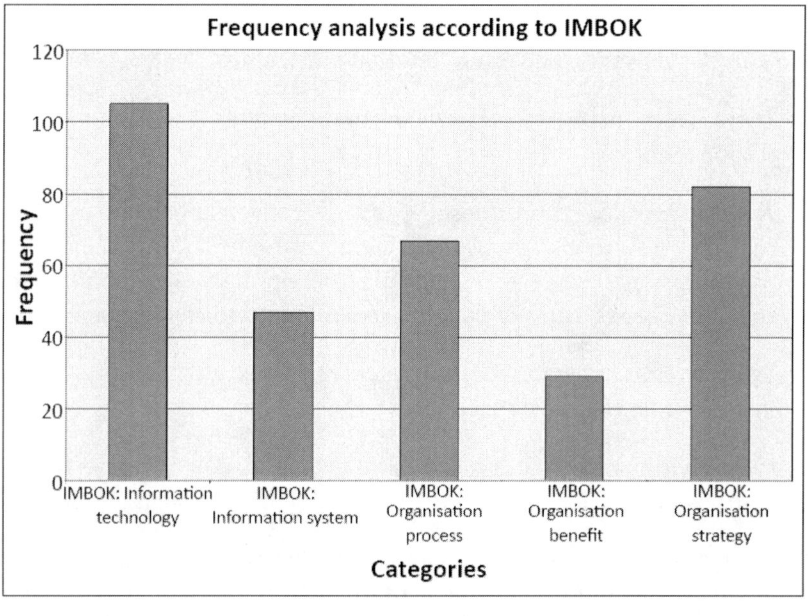

FIGURE 3: Category distribution by frequency.

- As already noted, the most frequently occurring discourse concerned *information technology*. This (it can be argued) is inevitable, when conversation is predicated on the emergence of ICTs in education, but it cautions us that much conversation might be grounded in the availability and enticing capabilities of information technology rather than simple educational need – we come back yet again to issues of technology 'push' and 'pull'.

- *Information systems* did not figure greatly in the conversations; the very concept of an 'information system' or 'application' was almost completely absent. This contrasts with business and public sector working, when we frequently see references to the 'Payroll', 'ERP (enterprise resource planning)', 'CRM (customer relations management)' and other typical 'applications' that are to be found therein. When information systems *are* referred to in the conversations, the references are generally rather weak except (in just one case) when referring to the 'learning management system' (LMS).

- *Organisational processes* were discussed more often, partly because in conversations there is usually a bias towards 'what I do and how', rather than 'why'; there is some interesting evidence within this process discourse about how ICTs are affecting and changing what happens in education.

- The *benefits of ICTs* were infrequently referenced and were often tangential, being implied rather than being explicitly articulated. This helps us to understand a major part of the problem, because if the organisation at large does not perceive or understand (and then manage) the benefits, then all the benefits are as good as lost. A strong discourse on the intended benefits of an ICT investment is essential if managers are going to have clear targets to aim for.

- The *strategy* discourse emerged weakly, in different ways and from different perspectives, and once again strategic issues were mostly implied rather than clearly articulated. However, this indicates that the seeds of strategic thinking are there, and in fairness to all in education we already know that change can take a long time and is never easy.

The paragraphs that follow present selected evidence that exemplifies some of the stronger evidence that is to be found in the conversations.

Who are the stakeholders?

These eight wide-ranging conversations were not structured in any particular way, and it is inevitable that the evidence turns around the role of teachers, learners and management. However, there is evidence of other critical stakeholders, and at the end of the day we have to remind ourselves that success comes from a balancing of all stakeholder expectations and an investment in resources and processes that will serve those expectations in a balanced way.

In the analysis of the conversations, there is evidence of 13 stakeholder groups, as indicated in Figure 4; of course, the frequencies of reference are not high enough, nor is the sample of respondents necessarily wide enough, to put a high level of significance on this result. Nevertheless, the analysis in the figure provides a clue as to which stakeholder groups might be the most important to consider, in the opinion of the respondents. And these stakeholder groups are not confined to teachers and learners, or students and professors; it is found that managers, communities, parents, administrators, other schools, suppliers and government are all factors in managing for success. A further investigation, in more detail, might reveal an even wider range of 'interested parties' and a better understanding of their relative importance, but we must assume that anyone, in any one of these stakeholder groups, could impact adversely on education in one or many educational institutions. When it comes to the adoption and integration of information technologies, it has been strongly argued that success will only emerge when the needs and expectations of all stakeholders are accommodated in the planning and implementation (Edwards & Peppard 1997).

Thus, these are the stakeholder groups that need to be understood, whose expectations will set the criteria of success or failure and who might in their own way impact on the efforts of educators and educational managers to deliver the best results. Their expectations are not only defined in terms of the benefits that they recognise, but also defined by the benefits that they seek.

DISCUSSION

Insights into the way that information management can affect the outcome of our investments in educational technology and our efforts to improve and advance our educational processes, and deliver strategic educational benefits, all come from a more detailed inspection of what was said by this diverse group of educators and administrators. The paragraphs that follow present selected evidence from the conversations, within the five domains of the IMBOK.

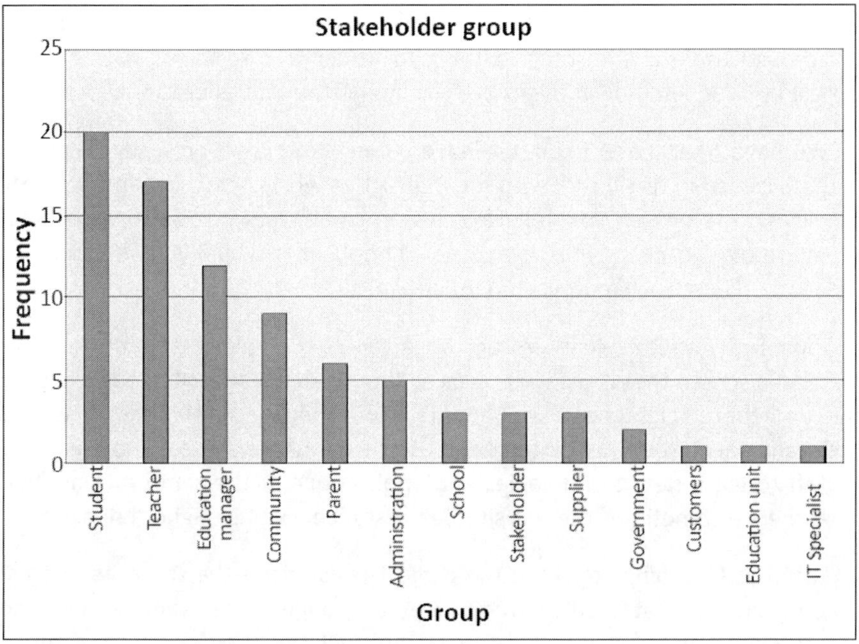

FIGURE 4: Stakeholder analysis.

TECHNOLOGY

As we have seen, there is a range of technology that was talked about, but the educational benefits that technology brings were often more implied than expressed. Computers dominated the conversations and even concern about the variety of brands in use was evident. The systems within a brand (and the 'trendiness' of the systems) were also evident in terms of obsolescence and the preference for new rather than upgraded technologies:

'We've just introduced the iPads - we have six iPads, three of them are the new iPad 2, they came today ... at this time we use the 'blackboard' app where they can write their words or their sentences as if with chalk, rather than have a traditional slate in front of them'. [Teacher, Kindergarten, Australia]

The replacement of the low-cost chalkboard with a very expensive continuously evolving tablet technology might not be the first choice of a developing country. The 'electronic' or 'interactive' whiteboard is one example that is becoming more common and does seem to bring benefits, although there is still some learning to be done in terms of how it might best be used and how it might be fully integrated into educational practice:

'We have been networked, we have connected classrooms, which means that I can sit in the library with the interactive white board on, and have the television screen on, and talk to a class in [a different place altogether and] say [*to our learners*] "here is another school, what would you like to say to them?"' [Teacher, Kindergarten, Australia]

Apparently, today's learners are so attuned to looking at (and learning from) a screen that it really appeals to them – even 'special needs' children – and there might be clear benefits from building a relationship with a distant peer group, as hinted here. But it is not always easy for teaching staff to learn new technologies and applications at the same rate as their learners – sometimes there is simple resistance, even in higher education:

'Faculties [*teaching staff*] are the biggest resistance - the students' biggest complaint is that faculty won't use it enough: "we want to use the technology and you won't". The faculty will say they don't have time'. [Learning Chair, University, Canada]

Sometimes it all seems *much* too challenging:

'Faculty need to be able to shift their way of thinking about the learning experience. So that's one part of that shift from the teaching paradigm - deliver information - to the learning paradigm ... so I see it as the duty of faculty to bring all these tools into the learning experience because it's like an apprenticeship process'. [Senior academic, University, USA]

Increasingly, students at higher education institutions choose (or are expected) to work with their laptops and not with pen and paper. One consequence is a whole additional layer of technology that requires a laboratory to be 'locked down' for an examination:

'We do some massive online testing. 220 students yesterday doing essay online in a lock-down exam, at the same time, in five labs ... we do a lot of lock-down, we actually use software to lock down the lab computers or even the student's own computers so they only have access to Blackboard, Turn-it-in. Some faculty adopt it, some don't'. [Learning Chair, Business school, Canada]

The complications of choosing to use technology accumulate as we do so. But at the institutional level some technologies are so basic that they are a 'must-have', otherwise students would simply choose to go to another institution with better facilities. Of course, understanding exactly which technologies qualify as 'important' is a strategic issue, not a technological issue. And between the technology and the achievement of strategic objectives come the information systems (or 'applications') that technology enables, and the processes that are enhanced by the systems.

Information systems

The more information-intensive areas of education showed very clearly how information systems are helping educators deal with large numbers of students and massive numbers of documents:

'Next term we will have nearly 2000 students in Business 101 and 102, and these are all being taught in collaborative manner. So there are 20 tables of 6 students each, every hour, dealing with business problems - the students submit one paper and review six within three days, and that involves 12 000 virtual copies of these papers going around and none getting lost'. [Learning management system manager, Business faculty, New Zealand]

Are there benefits from systems that handle such large numbers of students in this way? Probably, but detailed evidence of information systems that assist in education activities was actually limited. The balance of investment between administration and academic systems is a point of concern, and there is a tendency to invest in administrative systems first, although the connections with academic activity might be clear in specific cases:

'I actually did an administrative systems strategy for [*the university*] and they implemented most of it ... the area that we spent a lot of time on is academic support, administration of course management systems, and content'. [Learning Chair, Business school, Canada]

Assessment of learning figured strongly in the conversations. Traditional modes of teaching, learning and assessment are changed by the use of technology. For example, traditional assessment by means of tests and examinations can be substituted by the use of 'discussion boards', another example of an information system:

'I teach a class on business and 15% of the evaluation is based on the discussion board. I post a topic every week and I rate them on the number of posts, the quality of posts. I also rate them on helping one another, and it's got to be timely'. [Learning Chair, Business school, Canada]

Of course, systems merely augment the propensities of their users: if people like to live with chaos then they can create even greater chaos; if they like to be organised, then they can be even more organised. If they like to keep things and not throw them away, they can keep more things for longer:

'I was asked on the ferry this morning 'Sir, are you the "father" of [*the learning management system*]?' and I said 'Well, one of them!' and he said 'I just came back to the university, I've been away five years and when I logged in everything is still there! All the classes were there that I took before, all my notes, all my stuff, that's an amazing system!' I said "that's the way we designed it"'. [Learning management system manager, Business faculty, New Zealand]

For a university that wants to distinguish itself, and stand out as a great place to work as well as to learn, this is the sort of thing that we are looking for. But, it's a pity when systems fail, in this case one that was not developed in-house but by 'experts':

'The [*bought-in*] system was absolutely airtight up until recently, however the output [*from the new version*] is not time-stamped and depending on the vagaries of the local network and so on you could get an enrolment before a deletion, or a deletion before enrolment, or the system could wind up just deleting the people out of the course and nobody knew where they went ... This was a $20M investment, hundreds of people involved, some of them making more than $100k per year, and it looks like we have to write our own middleware!'. [Learning management system manager, Business faculty, New Zealand]

The conversations highlighted the reality that fundamental systems-related changes are at hand. In discussing plagiarism, and what to do about it, Google and Wikipedia quickly come up in conversation, but do they count as 'systems'? Are 'Google docs' (an example of 'software-as-a-service') to be seen as 'systems'? We are reminded that we do not have to do everything for ourselves any more. Web-based services, combined with the myriads of 'apps' for iPads and Android tablets, guarantee that we will increasingly reach outside our boundaries in order to find the systems and services that we need, in open markets and at very low cost. But not all 'open' applications are small, and the tension between commercially available and open source systems is palpable:

'In 2009 we completed the transition [*to Sakai*] ... I think it was a wise investment. But that's an example where you have to convince people in leadership positions that you are doing the right thing. For the longest time they said, "you're gonna use open source? What are you gonna do when it breaks?" My answer is: "look at these other schools, we are in a community"'. [Senior academic, University, USA]

The use of learning management systems was a recurring feature of the conversations, but what about other important educational activity, such as research? The substance and mode of research working is changing rapidly. Where a doctoral thesis might once have had a few photographs or drawings pasted in on blank pages, or bound in at the end, today we expect the all-singing all-dancing presentation of everything (that we made move, or sing or dance) in our research to be immediately available in the digital archive. Taken overall, the effort involved in achieving technology-induced change is keenly felt:

'So in a sense, by divesting some technology we take on new responsibility to guide and control the process and train people. And so we still need to look at all the assets that we have to make this happen, but to me I think it's a renaissance period for us, it's massive change, but it's so fun! It's like being on one of those roller coasters, you know? With the economic drag we felt like we were down here, but now we are going up again! It's really exciting, and a little scary'.[Senior academic, University, USA]

Clearly, this respondent is an optimist and an enthusiast. When it comes to changing the processes of education (dare we say 're-engineering' the processes?) things are not all that simple.

Process

Prevailing conventional education paradigms show signs of the impact of a developing technological world. Thus, knowing what goes on in a particular context is an important pre-requisite to understanding where an investment in technology might help. Tragically, the conversations reveal that learners are running ahead of many of the teachers, and teachers' control of teaching and learning processes is thereby threatened. Certainly the pace of change is a challenge that is more easily met by the learners (who have no 'baggage' to carry) than by the teachers, who are sometimes hidebound by their baggage. But when teachers choose to rise to the challenge there are rewards: 'It's amazing that the teachers create the content. It's the teachers that teach it so they are the ones to know what they want and what's good to learn' [Teacher, Kindergarten, Australia].

Not true of all, no doubt, but the liberation that comes from developing one's own material, as distinct from being told to teach from the 'standard' book that is chosen by some kind of educational authority, is a worthwhile reward for all positively oriented teachers, provided that they have the competency to deal with the pedagogical challenges that arise in designing and developing education programmes of different kinds.

New educational designs were evident in the high school in New Zealand:

'We've got a different setup in our school because we have what they call an open forum. There is no classroom. You teach in what we call a learning commons. What happens is, in a commons you can have four classes running at the same time. A common would be ... about the same size as this house. A class would be this size, so you would be here, and there next door (points to the kitchen) there would be another one. You can see them, hear them'. [Technology teacher, High school, New Zealand]

This was backed up by innovative use of cloud services, which are now an essential feature of learning at this school:

'I said to them, you are going to go into groups of five, the group leader is going to set up a Google doc and give you all editor rights, including me. The group leader is going to assign each person two sections and you are going to answer all the questions in your section. It worked quite well - you get a few stupid comments ... [*but*] one of the better students was a group leader and he basically blocked out two of the students in his group'. [Technology teacher, High school, New Zealand]

The use of Google Docs™ has since become more prevalent and the benefits (and other consequences) are becoming clearer. It is part of a move to make much more use of the social web in education. Close analysis of the evidence suggests a growing divergence between the formal and social aspects of educational processes; we must be aware of all the issues, and we must manage the change that will ensue in order to assure the benefits.

Benefits

We need to understand and communicate the benefits that are within our grasp, because it is the benefits that engage stakeholders when change is needed. Business models are changing dramatically and education is no different. But the whole question of strategic management in education, so important to the successful achievement of change, still seems to lie under the surface. Education is not immune to change, and has to be able to manage it. The conversations indicate that there is evidence of planning, but there is equal evidence of happenchance and improvisation. The following anecdote illustrates the need for strategic thinking based on understanding the benefits:

'One of the reasons we [*switched to Sakai*] is that there is an open source community of users that support the use of Sakai. Like other schools we seriously questioned whether Blackboard is the right choice at this point because it costs an awful lot of money and that was killing us; we asked how do we get support from Blackboard? Well it turns out that Blackboard had created a community of users that you could share your problems with, they get back to you and help. But I'm, like, 'Isn't that open source? That's open source isn't it? But we're paying someone to do this?' We don't want to do that, so we switched [*to Sakai*] in 2007'. [Senior academic, University, USA]

There are long-term operational and financial implications in decisions like these. But change is not just operational, it is fundamental; how we see knowledge and our perceptions of managing knowledge are changing, in very interesting ways:

'If something is classed as research, then one would assume that it is done in a rigorous manner and accepted by a body of scholars who are all agreed, and so if I presented a hypermedia [*work*] of, say, the structure of knowledge, it could be a taxonomy, some kind of graphical representation of the linkages of concepts and so on. If I found that these were accepted and integrated into the minds and work of other scholars, then the fact that I had rendered this representation of knowledge ... should be an even more compelling argument that I am a member of the research community ... and so this might be a shorthand way of avoiding writing thousands of pages'. [Learning management system manager, Business faculty, University, New Zealand]

This short monologue has profound implications. At one level we see people happily playing with, or even working with, 'mind maps'. But a mind map is just one example of how technology and systems allow us to deal with what we know in new and quite different ways. As the world still struggles to normalise and standardise the relatively simple thing that is metadata (such as the descriptive data about academic papers and books: author names, journal titles, publishers and so on) we have little prospect of normalising and standardising what knowledge is; but we are already a long, long way from the Dewey decimal classification system, and with these portentous issues in mind we have to agree that this all becomes highly strategic.

In summary, there is useful evidence of the potential benefits of information technology and information management here, but grasping them requires strategies that might redefine what education actually is.

Strategy

The stimulus that comes from new technologies drives much strategic change, in organisations of all kinds. Business, government and civil society all benefit from new ways of working and communicating. The answer, we are told, is to align our technology and organisational strategies:

[*Luftman*] is an American academic, former consultant. He did some quite interesting work on alignment ... [*but*] he has found 'the least aligned industry is education'! The reason is that there are multiple stakeholders with frequently conflicting goals, whereas in other businesses there are not. [*In education*] it's administration versus academics, academics versus students, and students versus administration. [Learning Chair, University, Canada]

Here is interesting evidence (Luftman 2003) that education really is late in understanding and adopting strategic management, but is strategic alignment 'doable' in education? Academic management is not universally known for its ability to manage everything well, and academics allow themselves the luxury (or is it a privilege?) to *do what they want to do*, much of the time. We have found that there are significant benefits to be gained, but tightening up the management of the many dependencies between technology acquisition and strategic success is what is needed, and this is indeed challenging.

At the heart of the confusion involved in aligning technology and strategy is the proper management of information, and of the information-enabled processes that we undertake to deliver educational value.

Results

The selected evidence presented here is from eight conversations and is therefore hardly definitive, but it helps us to adjust our thinking and achieve a balance. There is evidence of a range of stakeholders, and the ways in which they can help or hinder. There is evidence of a range of benefits that are sought, and we begin to see different kinds of benefit for the different stakeholders. Finally, even in these developed countries we see only passing signs of true strategic planning, but some evidence that planning does lead to better outcomes.

We are left with the most important question: *does strategic planning help to deliver the benefits of information management in education*?

It seems to depend on circumstances. Despite a concern for the ability of educational institutions to manage their technologies well, there is evidence that when educational organisations do get their strategic act together there are real benefits: for the teachers, for the learners and for other stakeholders. However, it is clear that the circumstances of the institution determine the nature of the strategy or strategies adopted.

Strategies for innovation, radical change and operational excellence

A closer examination of the specific evidence about strategies reveals thinking at three difference levels:

- A desire to innovate and discover new opportunities
- A commitment to undertake radical organisational change
- A simple ambition to work efficiently and effectively.

The high school in New Zealand is a good example of innovation. It is a new school with no baggage from its history that it has to deal with: it has new buildings, new ideas and new resources. Most importantly, its management meets regularly and drives the further development and refinement of the ideas by managing 're-invention':

One of the good things about the school is that it is totally new ... it's quite revolutionary! In the average school if something doesn't work people turn around and say 'these new ideas, new systems do not work ...' and go back to the old way ... Whereas in our case we have a strong tendency to say 'OK, it doesn't work, we have to re-invent ourselves'. The old way is not physically available, so you have to say 'right, let's re-invent ourselves'. [Technology teacher, High school, New Zealand]

This is real innovation, untrammelled by the baggage of a history.

The university in Auckland reveals quite a long history of careful and progressive management of its information technology, and although there are some signs of traditionalist thinking, the business school that contributed to these conversations has magnificent buildings and luxurious facilities within them. They seem to have close associations with business that keep their feet on the ground, and the management imperative probably rubs off when they meet and talk to their business friends. They are certainly making strategic use of the latest technologies:

There is a thing called 'infopath' which is building what appears to be a document but is actually a series of attributes that can be queried and mailed - this goes as input to Sharepoint and so basically common stuff – course description, course reports – all that kind of thing now sits in a repository in Sharepoint and it's versioned, and we can look at and across all of the documentation that used to be hard copy, and start to understand what it is that we can do with it. And [*yet*] we don't have integrated systems. Or the systems we have are not integrated enough. [Learning management system manager, Business faculty, University, New Zealand]

This looks like a strategically capable university that has a long history of proactively managing radical change (in relation to traditional university practice).

The university in the United States adopts a very traditional stance, but is highly aware of trends and has active programmes in place to consider its options and maintain its competitive standing (and, make no mistake, it is in a *very*competitive situation):

I think the challenge may be the senior leadership are always having to face those [*competitive*] issues and prioritise ... especially at a time of economic strain. Recently we did two successive rounds of cuts so, you see we are not a public institution and yet we are constrained by financial changes in the market. We rely on an endowment for certain spending. [Senior academic, University, USA]

Almost all of the strategic management effort in this institution is directed at operational efficiency, and maintaining the critical services that good students expect.

And so we find evidence of 'innovative', 'radical' and 'operational' strategies.

In the evidence, there are 82 instances of conversation that were coded as having a direct link to strategy; these are subdivided into the three subdivisions of strategy (see Figure 5). It is found that the majority of the instances reveal an operational approach to strategy (efficiency and effectiveness), rather than being truly strategic (grasping opportunities and managing them proactively) or innovative (making new opportunities through experimentation and invention)

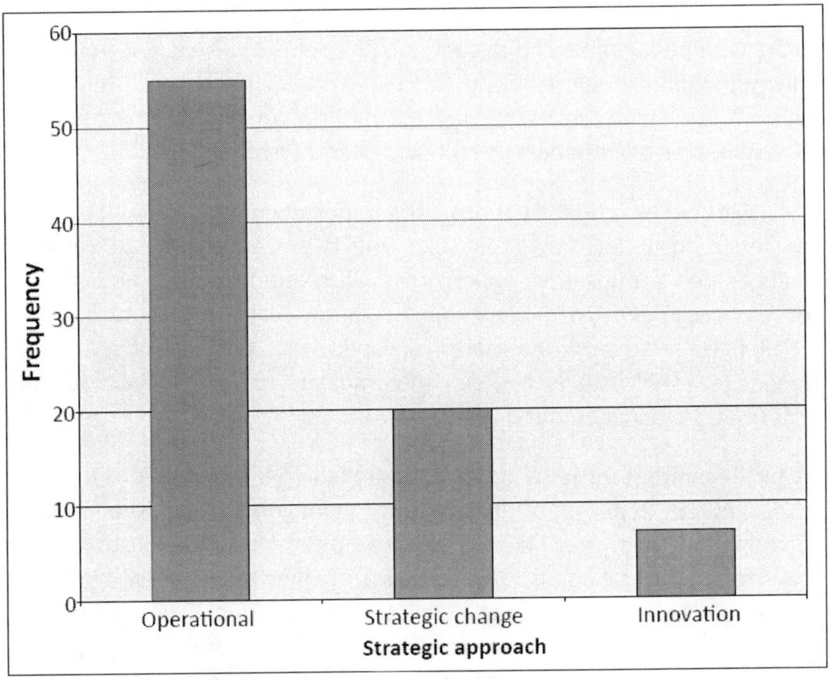

FIGURE 5: Analysis of management.

One might conclude that any educational institution that is willing and able to *innovate* will easily stand out from the crowd and ease itself much more successfully into the future, provided it can manage the innovation (and the consequential changes) well. Those that cannot innovate must focus on what is possible: being *distinguished* by doing some things very well and in different ways, or being *economical* by means of excellent administration and low-cost educational services. Technology can contribute to any of these, but doing more than one at the same time is confusing for staff and for students.

Implications of the work

This work began with the observation that education in South Africa is in a poor condition, and that there has been substantial investment in information technology that was intended to improve the quality of education, with variable results. Rather than looking at the problem from the 'bottom up', as often happens when looking at the impact of technology on pedagogy and didactics, it was decided to look at it from the 'top down', with a management perspective, and borrowing ideas from the domain of regular business management, and looking at information management broadly, not only as technology and systems:

- International experience confirms the importance of stakeholders, and the need to understand their roles and their expectations. Different stakeholder groups are looking for different benefits, and these differences need to be managed. South Africa presents special challenges relative to international experience, because of issues of language, ethnicity and context, thus making a strategic approach to information management doubly important.

- A further important issue, related to stakeholder expectations, is that the derived benefits of information management can be simple (leading to improved efficiency), complex (requiring institutional change in the cause of effectiveness) or innovative (new kinds of teaching and learning through evolutionary thinking). These different kinds of benefit cannot be achieved in parallel, they are conflicting. It is necessary to manage strategically by admitting and negotiating the benefits that are most appropriate to a situation. In most cases (from the conversations), the expected benefits were operational; in South Africa that might not be enough and more radical ideas might be needed.

- Perhaps most important to South Africa is that boundaries are dissolving, and education is becoming an international phenomenon. That might be of little interest to rural South African learners who are stuck in the traditional educational paradigm, but it is exacerbating the divide between the 'haves' and the 'have-nots'.

CONCLUSION

What can be seen from this analysis of international conversations, perhaps predictably, is that technology is both a threat and an opportunity. But there is evidence that education is – through the introduction of technology – becoming a very international business. This challenges every country to take stock of what it needs to do, where its educational capability and resources might come from and how they might be managed.

So will strategic planning help to deliver the benefits of information technology investments through good information management? We do not know; it is not yet clear that this is the case, despite the evidence found here. As we proceed further into the digital age, we need to be conscious that:

- ICT itself is, first and foremost, a *cost driver* not a *benefit driver*: benefits come from improved processes that deliver improved organisational performance. But education is increasingly recognised as an incremental process that is driven by the individual learner, not by the institution.

- It is necessary to be certain about (and committed to) the delivery of *benefits to all stakeholders*, not just teachers and learners, but the tensions between stakeholders are real and impede a simple agreement about what the benefits actually are, as argued by Luftman (2000, 2003). Truly strategic thinking sets targets, manages demand and supply and ensures that all our investments in ICTs are directed at real benefits, for real stakeholders.

- Education systems are a part of a global ecology that is rapidly *eliminating geographical and cultural boundaries*, with real consequences for the prospects of our young populations. Even the best strategy managers struggle to deal with the huge gulf of understanding that can arise when cultures clash.

Life in education is full of surprises and so not everything can be planned. Improvisation is the stuff of innovation, and so teaching practice must not be constrained. But the *balance* between planning and improvisation must be understood and managed. What cannot be tolerated in countries and regions where education is desperately needed is the chaotic and anarchistic behaviour that might result if the adoption of ICTs in education is not well managed, in ways to deliver real benefits to the most important stakeholder: the learner.

ACKNOWLEDGEMENTS

Competing interests

The authors declare that they have no financial or personal relationship(s) that may have inappropriately influenced them in writing this article.

AUTHORS' CONTRIBUTIONS

A.B. (University of the Western Cape) undertook research design and data collection, and drafted the article. I.M.V. (University of the Western Cape) mentored and contributed significantly to the finalisation of the article.

REFERENCES

1. Anderson, J., Van Weert, T., Buettner, Y., Duchâteau, C. & Fulford, C., 2002, *Information and communication technology: A curriculum for schools and programme of teacher development*, UNESCO, France.

2. Angell, I., 2000, *The new barbarian manifesto: How to survive the information age*, Kogan Page, London, England.

3. Baets, W., 1992, 'Aligning information systems with business strategy', *The Journal of Strategic Information Systems* 1(4), 205–213.

4. Baker-Eveleth, L., Eveleth, D.M., O'Neill, M. & Stone, R.W., 2007, 'Enabling laptop exams using secure software: Applying the technology acceptance model', *Journal of Information Systems Education* 78, 413–420.

5. Barron, A.E., Kemker, K., Harmes, C. & Kalaydijian, K., 2003, Large-scale research study on technology in K- 12 school: Technology integration as it relates to national technology standards', *Journal of Research on Technology in Education* 35, 489–508.

6. Bhusry, M. & Ranjan, J., 2012, 'Enhancing the teaching-learning process: A knowledge management approach',*International Journal of Educational Management* 26(3), 313–329.

7. Botha, N., 2009, 'Analysis and summary of the research themes of the concept papers received in response to the two national calls from the NRF, National Research Foundation', viewed 14 August 2010, fromhttp://www.nrf.ac.za/files/Document%203-Summary%20research%20themes%20national%20research%20agenda%20June%2009%20_2_.pdf

8. Brown, C., 2010, 'Excavating the meaning of information and communication technology use amongst South African university students: A critical discourse analysis', doctoral thesis, Dept. of Information Systems, University of Cape Town.

9. Bytheway, A., 1998, 'Academic CS/IS Infrastructure in South Africa: An exploratory stakeholder perspective, *Proceedings of SAICSIT* November 1998', Cape Town, South Africa.

10. Bytheway, A., 2004, *ICT in Higher Education research project*, University of the Western Cape, Bellville.

11. Bytheway, A., 2004, 'Information management body of knowledge (IMBOK)', Cape Technikon, Cape Town.

12. Bytheway, A., Sadeck, O., Dumas, C., Chigona, W., Chigona, A., Mooketsi, B. & Fanni, F., 2010, 'Integrating ICTs into the classroom: Assisting teachers in disadvantaged primary schools', *Proc 2010 eSkills Summit*, Cape Town, South Africa.

13. Bytheway, A., 2011, 'Assessing Information Management Competencies in Organisations', *Electronic Journal of Information Systems Evaluation* 14(2), 179–192.

14. Bytheway, A., 2013, 'Qualitative Research without money: Experiences with a home-grown Qualitative Content Analysis tool', *The Journal of Community Informatics* 9(4).

15. Chatham, R. & Patching, K., 2000, *Corporate politics for IT managers: How to get streetwise*, Butterworth-Heinemann, Newton.

16. Chatterji, S., 2007, 'Bridging business and IT strategies with enterprise architecture: Realising the real value of business-IT alignment', *Information Systems Control Journal* 3, 1–2.

17. Chigona, W., Bladergroen, M., Cox, S., Dumas, C. & Van Zyl, I., 2011, 'Educator discourses on ICT in Education', in *E-Skilling for Equitable Prosperity and Global Competiveness*, South African Department of Communications, East London, South Africa.

18. Davids, Z., 2009, 'The educators' perspective on the factors that influence the success of ICT school initiatives within the Western Cape', master's thesis, University of Cape Town, South Africa.

19. DeLone, W.H. & McLean, E.R., 1992, Information systems success: The quest for the dependent variable', *Information Systems Research* 3(1), 60–95.

20. DeLone, W.H. & McLean, E.R., 2003, 'The DeLone and McLean model of information systems success: A ten-year update',*Journal of Management Information Systems* 19(4), 9–30.

21. Edwards, C. & Peppard, J., 1997, 'Operationalizing strategy through process', *Long Range Planning* 30(5), 753–767.

22. Hardman, J. & Paucar-Caceres, A., 2010, A soft systems methodology (SSM) based framework for evaluating managed learning environments', *Systemic Practice and Action Research* 24(2), 165–185.

23. Henderson, J.C. & Venkatraman, N., 1993, Strategic alignment: Leveraging information technology for transforming organizations', *IBM Systems Journal* 32(1), 4–16.

24. Jansen, J., 2012, 'Seven costly mistakes', *Times Live*, viewed n.d., fromhttp://www.timeslive.co.za/opinion/columnists/2012/04/05/seven-costly-mistakes

25. Kearns, G.S. & Lederer, A.L., 2000, 'The effect of strategic alignment on the use of IS-based resources for competitive advantage', *The Journal of Strategic Information Systems* 9(4), 265–293.

26. Knights, D. & Murray, F., 1994, *Managers divided*, John Wiley, Chichester.

27. Koch, A., 2006, 'A conceptual model for a Co-operative Education Management system for tertiary institutions in South Africa', doctoral thesis, Cape Peninsula University of Technology, Cape Town, South Africa.

28. 'Labour's computer blunders cost £26,326 bn', The London Independent, viewed 24 August 2010, fromhttp://www.independent.co.uk

29. Luftman, J., 2000, 'Assessing business-IT alignment maturity', *Communications of the Association for Information Systems*4, 1–51.

30. Luftman, J., 2003, 'Assessing IT/business alignment', *Information Systems Management* 20(4), 9–15.

31. Madiba, M., 2009, 'Investigating design issues in e-learning', doctoral thesis, Dept. of Information Systems, University of the Western Cape, Bellville, South Africa.

32. Maslin, N., 1990, *The intelligent island*, television broadcast, BBC, London.

33. Mason, R.O. & Mitroff, I.I., 1973, 'A program for research on management information systems', *Management Science*19(5), 475–487.

34. McFarlan, F.W., 1984, 'Information technology changes the way you compete', *Harvard Business Review* 62(3), 98–103.

35. Michael, S.O., 1998, 'Best practices in information technology (IT) management: insights from K-12 schools' technology audits', *International Journal of Educational Management* 12(6), 277–288.

36. Moloi, K., 2007, 'An overview of education management in South Africa', *South African Journal of Education* 27(3), 463–476.

37. Motala, E., 2009, *Guidelines informing the NRF's choices and priorities for Education Research Funding: A report prepared pursuant to the NRF's regional workshops held during October and November 2009*, NRF, Pretoria.

38. Omona, W., Van der Weide, T. & Lubega, J., 2010, 'Using ICT to enhance knowledge management in higher education: A conceptual framework and research agenda', *International Journal of Education and Development Using ICT* 6(4), 83–101.

39. Rahimi, F., 2010, 'The role of information and communication technology (ICT) in a higher education institution: With specific reference to disadvantaged students, cultural aspects and motivation', doctoral thesis, University of Pretoria, South Africa.

40. Republic of South Africa, 2004, *White paper on education*, Government Printers, Pretoria, South Africa.

41. Shirky, C., 2008, *Here comes everybody: the power of organizing without organizations*, Penguin, New York.

42. Strauss, A.L. & Corbin, J.M., 1998, *Basics of qualitative research: Techniques and procedures for developing grounded theory*, Sage Publications, Thousand Oaks.

43. Uys, P., 2007, 'Enterprise-wide technological transformation in higher education: The LASO model', *International Journal of Educational Management* 21(3), 238–253.

44. Venkatesh, V., Morris, M.G., Davis, G.B. & Davis, F., 2003, 'User acceptance of information technology: Toward a unified view', *MIS Quarterly* 27, 425–478.

45. Ward, J. & Daniel, E., 2005, *Benefits management: Delivering value from IS and IT investments*, Wiley, Chichester.

46. Ward, J. & Peppard, J., 2002, *Strategic planning for information systems*, 3rd edn., Wiley, Chichester.

47. Zachman, J.A., 1987, 'A framework for information systems architecture', *IBM Systems Journal* 26(3), 590–616.

Chapter 2

Electronic Resource Management Systems

Mark Ellingsen

ABSTRACT

Computer applications which deal with electronic resource management (ERM) are quite a recent development. They have grown out of the need to manage the burgeoning number of electronic resources particularly electronic journals. Typically, in the early years of e-journal acquisition, library staff provided an easy means of accessing these journals by providing an alphabetical list on a web page. Some went as far as categorising the e-journals by subject and then grouping the journals either on a single web page or by using multiple pages. It didn't take long before it was recognised that it would be more efficient to dynamically generate the pages from a database rather than to continually edit the pages manually. Of course, once the descriptive metadata for an electronic journal was held within a database the next logical step was to provide administrative forms whereby that metadata could be manipulated. This in turn led to demands for incorporating more information and more functionality into the developing application.

Before long, many institutions were devoting resources to developing and maintaining systems which could manage a range of electronic resources, including information regarding abstracting and indexing services as well as electronic journals. In early 2001, Tim Jewell of the University of Washington carried out asurvey for the Digital Library Federation (DLF) of such in-house developments in North American universities (Jewell, 2001). This showed that libraries were trying to present and maintain information regarding e-resources, which often focussed on particular subsets of this information but that - as one would expect - there were many common features across the different systems being developed. One of the most well known of these systems is the one developed by MIT called VERA(Virtual Electronic Resource Access).[1]However, these in-house developments were not confined to the US. For example, my own institution in the UK, the University of Bristol has also developed an ERM application, which allows staff to input metadata and users to search and browse titles.

What is interesting, however, is that these applications were developed in-house to respond to the lack of functionality within existing library management systems which handled the major part of library processes.[2]It has taken some time for library system vendors to catch up and one can now find a list on the ERM Web Hub, provided by Cornell University, although this does not claim to be comprehensive. However, to my knowledge, as yet there has been no comprehensive comparison of the functionality provided by these systems similar to the one produced by Tim Jewel for in-house developed applications. As systems mature this will become an important task to help libraries make an informed choice.

FUNCTIONALITY

In what follows I will indicate some of the functional areas that an ERM system should cover. In particular, I will emphasise some of the areas in which the functionality differs from that required by the management of print journals.

Like most other library resources, electronic resources need to be acquired in some way. However, the practice of acquiring and maintaining electronic journals differs in some important respects from the management of print journals. For example, libraries often acquire e-journals as part of a package of resources from an e-resource aggregator such as EBSCO. There is a need to determine which journals are covered by the package and for what period of time. In relation to that, users often need to know when a new issue is available. In the circumstance when there is no physical item sent to the library it is difficult for library staff to maintain this information. If libraries wish to keep track of that information then there is a need for some sort of electronic check-in process.

Access to the resource is another area where the electronic format differs from the printed version. Access to electronic resources is subject to the risk of network or host hardware disruption. There is a need to be able to flag the unavailability of an individual resource or a package of resources to the potential user early on, thus mitigating any frustration about the level of the service. Secondly, electronic resources are often licensed to the institution and this adds a further level of complexity, which is not attached to print journals. Furthermore, access is often restricted via network address or via some authentication and authorisation mechanism.

While the selection and evaluation of serials in the printed format is an important process, there are special considerations when it comes to the same process as applied to the electronic format. In particular, there are important elements to be recorded about the user interface such as the technical requirements for the evaluation and whether the local infrastructure can be configured to meet these criteria. For example, web browser compatibility and any associated plug-ins need to be recorded because this may have an impact on the rollout of browser configurations to library and faculty PCs or even whether the institution can support the interface at all. But it is not just technical considerations, which need to be taken into account. The usability of the interface needs to be evaluated and there may well be a choice of interfaces from different providers to the same package of resources or subsets thereof. The choice of interface needs to be justified to the faculties and the institution as a whole and the reasons for the choice need to be recorded.

Like print journals, e-journals need to be acquired and paid for, and the appropriate sums associated with the appropriate budgets. It is a fair assumption to make that the acquisitions module of the library management system should handle the recording of financial transactions at least while there is no standard interface to central institutional systems. However, as mentioned earlier, electronic journals are often bundled into packages. Although this is not unknown in the print world, the greater prevalence of e-journal packages necessitates a sophisticated package management process within the acquisition module. There needs to be a flexibility to have the option of binding the financial transaction to the whole package or to distribute portions of the transaction to the components of the packages, the individual titles. Furthermore, the print and electronic formats may be linked such that the cancellation of the print format may invalidate the license agreement for the electronic format. The system must be able to handle this link and embed the relationship within an appropriate workflow. There is also a need to handle situations where there are separate payments to the package licensor and to an interface or multiple interface providers. For example, one may pay Oxford University Press for the e-journal content but also pay both them and HighWire for the interfaces.

There are other aspects of administering the resource, which are specific to electronic resources. For example, some interface providers allow institutions to brand the interface so that users are aware that it is being paid for by that institution. Information about how the interface is branded should be kept within the ERM system so that electronic resource library staff can keep track of where branding occurs which would make it easier to make global changes. Consequently, the system would also need to record administrative accounts, i.e. usernames and passwords, to each interface provider where appropriate. Also, as mentioned earlier, it would be very useful if the e-resource librarian could flag whether an individual resource or a package of resources was not currently available across the network and to log periods of downtime. The latter would give the librarian some indication of the reliability of the resource and together with usage statistics could provide input into decision making as to whether to continue with the resource or interface provider. Of course, it is difficult to record the usage of resources particularly when the user moves away from locally administered web pages to an provider of a package of resources. More often than not library staff rely on statistics provided by the vendor and these statistics come in many formats. Recently, there have been moves to standardise usage statistics from vendors, of particular note in this field is COUNTER. There may be a requirement to upload COUNTER statistics into the ERM database so that usage statistics can easily be analysed together with other data pertaining to an e-resource.

Unlike print material, which does not usually come bundled with access restrictions, access to an electronic resource requires some form of authentication and authorisation. At the very least, an ERM system must be able to record information about the mechanisms used for authentication and be able to provide that information to users if necessary. One wouldn't expect it to handle the authentication mechanism itself as this would either be done at the host site or by some form of distributed authentication mechanism. However, what if only subsets of our users are authorised to use this resource? Again if we wish to present to users only those resources that they are authorised to use then it needs not be the ERM system which does that. Presentation of resources could be done via a portal and based upon user profiles to determine which resources the user was authorised to use. However, one would expect that the ERM system would provide some means of gathering all information pertinent to a resource and this may include which categories or groups of users were authorised to use it. It would be vital to see this information in one place if one were negotiating or re-negotiating agreements and contracts.

Electronic resources are increasingly governed by licensing terms and conditions. We expect the user to have read these or at least to be aware of the restrictions of use. Recording these restrictions within the ERM system would make it easier to present a summary of restrictions to the user. For example, whether the user can download material or use the items in course packs or whether the library staff can use the resource to satisfy interlending requests. Recording this information within an ERM database will allow libraries to display this information to users in a consistent manner.

These are just some of the functional areas that an e-resource management system should handle The functionality required by an electronic resource management system has evolved over time as library staff have become more acquainted with the processes involved with managing these resources and providers have gained experience in the provision of resources across the Internet.

THE DIGITAL LIBRARY ERM INITIATIVE

The Digital Library Federation have been very proactive in encouraging the development of standards in this area. It was within this context that Tim Jewel conducted a survey of the functionality offered by in-house developed ERM systems in North America. Together with Adam Chandler of Cornell University they organised a web site to act as focus for this information (Medeiros, 2003). A meeting was held at the ALA annual conference in June 2001 which led to the setting up of an informal steering group. This group presented a workshop on ERM standards at a meeting sponsored by the DLF and NISO in June 2002. The workshop was not only attended by librarians but by library system vendors and serials publishers. It was agreed that standards were a key element to ensure successful developments of ERM systems and to this end it was agreed to provide a more formal and collaborative organisation to this work. A more formal steering committee was formed as well as two reactor panels to provide expert advice. One panel was made up of librarians with an interest or experience in managing electronic resources, and the other was made up of library system vendors and serials publishers amongst others.

The initiative's aim is to provide the community with a set of specifications, which would encourage the development of electronic resource management systems, based on standards and best practices. To this end it has produced a number of concrete deliverables. It has produced an entity-relationship diagram, supported by a data dictionary and a description of data structures, which maps data elements to the entities involved in electronic resources as well as to map the relationship between these entities. Secondly, it has produced a functional requirements specification and a workflow diagram Finally, the initiative looked at the possibility of providing an XML (Extensible Markup Language) schema to encapsulate some of the ERM data elements. XML is a mark up language, which is designed amongst other things to describe data and facilitate its exchange. An XML schema is a definition of the constraints which an XML document must adhere to. XML and associated schema can facilitate data exchange between electronic resource providers and the library, as well as between the library and other systems such as course management systems. The initiative has produced a schema for encapsulating license data as a proof of concept and how that schema may relate to existing rights expression languages such as the Open Digital Rights Language (ODRL) Initiative and the Creative Commons RDF schema. In particular, example use cases have been provided to deal with the exchange of licensing information. All these deliverables are attached as appendices to the final report, which is now available (Jewell, 2004).

STANDARDS

An ERM system can be a module within an integrated library system or a stand-alone application. Whichever way the system is developed, it needs to be integrated not only with the ILS but with other applications and services. However, it should be noted that there is usually more than one way to integrate systems and it is not necessarily obvious which subset of data should reside in which application. The partition of data between an ILS, an ERM system, a link resolver database, e-resource aggregators and a LDAP (Lightweight Directory Access Protocol) directory service - to name a few data stores where information relevant to access to e-resources may be kept - is not a foregone conclusion. However, whichever architecture vendors opt for, application integration is made a whole lot easier if standards are adopted. In the following I would like to mention a few example to illustrate the possibilities and the importance of standards.

Some services may need to query an ERM system if the latter has its own database of e-journal metadata. Typically this could be done through a Z39.50 query or the newer Search/Retrieve Web Service (SRW) initiative based on web services standards. Metadata could also be exchanged between subscription agents and the ERM system. For example, the ONIX for Serials standard, developed by EDItEURand NISO, can facilitate the exchange of serials subscriptions or holdings. It may also facilitate automatic electronic check-in.

The metadata in the ERM database could be used as a source of an OpenURLmessage package, which could be used as a basis for context sensitive linking. Context sensitive linking is often discussed within the context of access to an item level resource such as a journal article. However, the OpenURL standard can also be used to generate a link to the journal as a whole. This often occurs if there is not enough information in the OpenURL package to generate a link to the specific item. But it may also be useful to provide an OpenURL for each member of a list of e-journals. That list could be generated from the ERM system using other protocols such as SRW or Z39.50.

However, there may be cases where an e-resource is only available to a particular community within the institution. Bringing together user profile information with e-resource metadata as an access profile may be the job of the ERM system with the appropriate hooks into a directory service via the LDAP protocol. The ERM system can then provide information to third party applications in terms of presenting links to the resources tailored to the profile of the user. It may also provide metadata to an authorisation mechanism based around user identities such as Shibboleth, though this is not the only way this can be done. A combination of group information held in an LDAP directory and access information held at the data provider's site may suffice. There is more than one way to build user profiles for authentication and authorisation. This information can also be provided to a link resolver so that it is passed on to a data provider via the OpenURL protocol. Version 1.0 of the OpenURL standard allows for the passing of user information to the resource.

There are other standards, which are useful for application integration, and one of the most important is the Simple Object Access Protocol (SOAP). SOAP is an XML based protocol, which facilitates the exchange of information between applications and the calling of procedures remotely between applications over HTTP. It is one of the core standards for the web services architecture for integrating applications built on heterogeneous platforms A second standard which is of particular relevance to the integration of applications within portals is the Web Services for Remote Portlets (WSRP) standard. This allows portal developers to plug-in remote presentation oriented web services as a portlet. In particular, the ERM vendors need to provide a web service built to the WSRP standard to allow the embedding of the service within a library or institutional portal. Of course, the portal application itself needs to conform to the WSRP standard in order to plug in WSRP services As a matter of course, ERM vendors should provide appropriate interfaces, some of which may be provided as a web service, to facilitate the integration with other applications. Those services which may need to communicate with third party applications are candidates for development under the web services framework. As mentioned earlier, web services may be used for presentation purposes as in WSRP portlets or to facilitate communication and exchange of data between applications.

So, the development of ERM systems must take into account current and emerging standards if they are to integrate with the portfolio of library applications. However, it is just as important that they can be embedded within integration frameworks such as portals. There has long been a demand from both public sector and corporate organisations that library applications be seen as one component in the delivery of services to users. The key to interoperability is the development of systems which conform to standards as well as having published APIs. The importance of the emergence of web services as a set of standards for application integration should make it easier to integrate library applications with other institutional systems. Library system vendors need to build in these standards into their products.

CONCLUSION

In this paper we have looked at the emergence of ERM systems, the functionality that is required of them, and the push to adopt standards in their development. Many of these systems are not yet mature and it will be interesting to see how they develop over the coming years. What is of particular interest is how the management of electronic books may be integrated within these systems. The difference in scale of the amount of material and the technological and functional complexity in managing e-book resources within the context of a virtual or hybrid library is a challenge that has yet to be met.

REFERENCES

1. Ebenezer, C.: "Trends in Integrated Library Systems" *Vine*, 32(2003)4, 19-45.http://dlist.sir.arizona.edu/archive/00000241/

2. Hennig, N.: "Improving Access to E-Journals and Databases at the MIT Libraries: Building a Database-Backed Web Site Called 'Vera'". *The Serials Librarian*, 41(2002)3-4, 227-254. http://www.hennigweb.com/publications/vera.html

3. Jewell,Timothy D. [et al.]. *Electronic Resource Management. The Report of the DLF Initiative*. Washington : Digital Library Federation, August 2004.http://www.diglib.org/pubs/dlfermi0408/dlfermi0408.pdf

4. Jewell, Tim: Online Acquisition Systems survey in North American Universities, 2001. http://www.library.cornell.edu/cts/elicensestudy/u-washington/FinalAppendixB.xls

5. Medeiros, N.:"A Pioneering Spirit: Using Administrative Metadata to Manage Electronic Resources". *OCLC Systems and Services*, 19(2003)3.http://www.library.cornell.edu/cts/elicensestudy/pioneering_spirit.pdf

6. WEB SITES REFERRED TO IN THE TEXT

7. COUNTER - **C**ounting **O**nline **U**sage of **N**etworked **E**lectronic **R**esources.http://www.projectcounter.org/

8. Creative Commons. http://creativecommons.org/

9. DLF - Digital Library Federation. http://www.diglib.org/dlfhomepage.htm

10. DLF ERM - Digital Library Federation Electronic Resource Management Initiative.http://www.diglib.org/standards/dlf-erm02.htm

11. EDItEUR. http://www.editeur.org/

12. HighWire Press. http://highwire.stanford.edu/

13. LDAP - Lightweight Directory Access Protocol. http://www.openldap.org/

14. NISO - National Information Standards Organization. http://www.niso.org/

15. NISO Committee AX- Development of an OpenURL Standard.http://library.caltech.edu/openurl/

16. ODRL - Open Digital Rights Language initiative. http://odrl.net/

17. ONIX for Serials. http://www.editeur.org/onixserials.html

18. OUP - Oxford University Press. http://www.oup.co.uk/

19. Shibboleth. http://shibboleth.internet2.edu/

20. SOAP - Simple Object Access Protocol. http://www.w3.org/TR/soap/

21. SRW - Search/Retrieve Web Service. http://www.loc.gov/z3950/agency/zing/srw/

22. University of Bristol Information Services. http://www.bris.ac.uk/is

23. VERA - Virtual Electronic Resource Access.http://libraries.mit.edu/vera

24. Web Hub for Developing Administrative Metadata for Electronic Resource Management. http://www.library.cornell.edu/cts/elicensestudy/home.html

25. Web Services Activity Statement. http://www.w3.org/2002/ws/Activity

26. WSRP - Web Services for Remote Portlets. http://www.oasis-open.org/committees/wsrp

27. XML - Extensible Markup Language. http://www.w3.org/XML/

28. Z39.50. http://www.loc.gov/z3950/agency/

NOTES

[1] On the development of VERA see Hennig (2002).

[2] For a paper on trends in library management systems see Ebenezer (2003).

Chapter 3

Developing an Electronic Resource Management System: Verde from Ex Libris

Tamar Sadeh

ABSTRACT

Rapid growth in the number of electronic resources and the complexity of managing e-collections has posed new challenges for libraries and hence requires the development of new tools. To meet this need, Ex Libris is developing a new product, an electronic resource management (ERM) system. The development process began in 2002 and has benefited from the company's 25 years of experience in providing libraries with various products - an integrated library system, digital asset management system, library portal, and link server.

Working with an international focus group and development partners and interacting with the Digital Library Federation's Electronic Resource Management Initiative (DLF ERMI) committee, Ex Libris explored the various aspects of the e-resource life cycle. As a result, the Ex Libris ERM system was designed to address the issues raised through this collaboration and the DLF requirements. The interface of the system represents all of its components; it includes navigation options and facilitates workflows that support the various activities that librarians perform when dealing with electronic resources.

This paper describes the development process of the Ex Libris ERM system. The emphasis is on the particular functionality required for managing e-resources and the ways in which existing systems in the library arena can handle specific tasks.

INTRODUCTION

Librarians often comment that colleagues who deal with electronic resources do not last long in this role. These librarians are facing a task that is becoming more and more difficult - but help is, indeed, on the way, as I intend to show here.

Several factors have rendered the job of such librarians extremely challenging: the amazing growth of electronic collections and the increasingly central role they are playing in libraries, the large budgets that are involved in acquiring these collections, the endless variation in the packages offered by the many publishers, interface providers and subscription agencies in the market , the frequent changes in business models, and, above all, the lack of automated tools to deal with the complexity of e-resource management.

As they attempt to maintain some control over their e-collections, librarians find themselves lost in a morass of spreadsheets and e-mail messages, and dealing with a variety of independent systems and data containers that are not integrated with each other. Too often librarians rely on their memory alone to co-ordinate systems such as the acquisition module of their integrated library system, their alphabetic lists of electronic journals and databases, a metasearch tool, and their local link server. In addition to the initial effort of setting up information in multiple places and the potential lack of consistency between systems, there might also be considerable duplication of effort. Much of a librarian's ability to carry out necessary tasks is dependent upon personal experience. Moreover, the knowledge and experience gained from dealing with e-resources is often vested in too few people - sometimes only one - a situation that leaves libraries exposed. Furthermore, providing meaningful metrics, such as detailed cost analyses and statistics on past performance and usage, remains problematic. Without such metrics, managers cannot make fully-informed decisions. With spending on e-collections increasing rapidly, a better solution is needed.

There is no doubt that libraries are ready for 'a system that supports management of the information and workflows necessary efficiently to select, evaluate, acquire, maintain and provide informed access to electronic resources in accordance with their business and license terms'

(Anderson, et al., 2004). A number of vendors are now working on providing such a solution. Some vendors will offer an ERM system as an integral part of their integrated library system, while for others - including Ex Libris - the system is a stand-alone module. The latter solution is increasingly the trend; in a recent article, North Carolina State University librarians Greg Raschke and Suzanne Weiner put forth a good case for such separation while emphasizing the need for tight integration (Raschke & Weiner, 2004).

Any new system needs to integrate well with the automated library environment, where existing systems already support various aspects of the e-resource life cycle. The new system should not duplicate existing data and procedures but rather complement them while streamlining the workflows; it should provide a central 'control tower' for librarians from which they can manage the e-resource environment.

WHAT CAN LIBRARIANS DO WITH AN E-RESOURCE MANAGEMENT SYSTEM THAT THEY CANNOT DO TODAY?

The first challenge is to define what is meant exactly by *e-resource*. An individual electronic journal such as D-Lib Magazine is an e-resource, as is an electronic book or an abstracting and indexing database such as MEDLINE; but an e-resource can also be a package of e-journals or a database of abstracts and indexes that includes the full text of some or all articles referenced by the indexes. Furthermore, we cannot consider an e-resource without the interface through which it is offered; these elements are intricately linked, although they can be licensed separately.

Let us look at an example of a complex e-resource, Oxford University Press (as of August 2004):

- Oxford University Press (OUP) publishes about 190 e-journals. A library can buy a package of these e-journals from the publisher itself; via an agent (for example, the NorthEast Research Libraries consortium (NERL) in the United States; or through various information vendors.

- OUP provides an interface for accessing many OUP e-journals; however, other providers, such as HighWire Press, Periodicals Service Company (PSC), and ProQuest offer their own interface to OUP journals as well.

- OUP offers subscriptions to all the journals that it publishes.
- Through the OUP interface, users can access the full text of only some OUP journals. In the OUP journal list many journals are linked instead to theHighWire Press interface.

  Although HighWire Press offers an interface that allows OUP journal subscribers to access many OUP journals, the OUP collection is not available for subscription through HighWire Press.[1] For some of the HighWire-hosted OUP journals, HighWire Press offers free full-text access subject to an embargo period.

Licensing an e-resource, therefore, is far from trivial. In addition to all the limitations just mentioned, an OUP package might be governed by a set of license terms, whereas a specific OUP journal - for example, *JNCI Cancer Spectrum* - might be governed by a different set of terms. This particular journal was included in the 2000-2001 package of Oxford journals; at the beginning of 2002, the journal became subject to a separate license addendum with different business terms that were no longer tied to print.

Other factors that are specific to e-resources and do not apply to the traditional print world include authentication, access, administration, usage, etc. These two types of resources differ in the manner in which they are acquired, accessed and licensed. An ERM system needs to support these differences by providing the infrastructure for both the data and the workflows that are characteristic of e-resources and are missing from systems dealing with print resources. The system should provide the tools to manage e-resources throughout their life-cycle while supplying the data that administrators need to make decisions about acquisitions, renewals and cancellations of e-resource subscriptions. Moreover, the system needs to be extremely flexible and open, capable of supporting rapidly-changing business models.

The electronic resource management system (ERM) of Ex Libris, Verde, was designed from the outset to address these issues.

LIFE CYCLE OF E-RESOURCES

One of the main goals of the Verde electronic resource management system is to support electronic collection development. Currently, for each resource that librarians want to acquire, they must learn which of many diverse sources will enable them to acquire the resource - and, needless to say, a librarian cannot always be aware of all the available options. Verde facilitates this process by providing a global picture of the e-resource world: a comprehensive description of what is available, including the information about the packages, the interfaces, and the various providers. This description is the point at which the resource's life-cycle in the library begins.

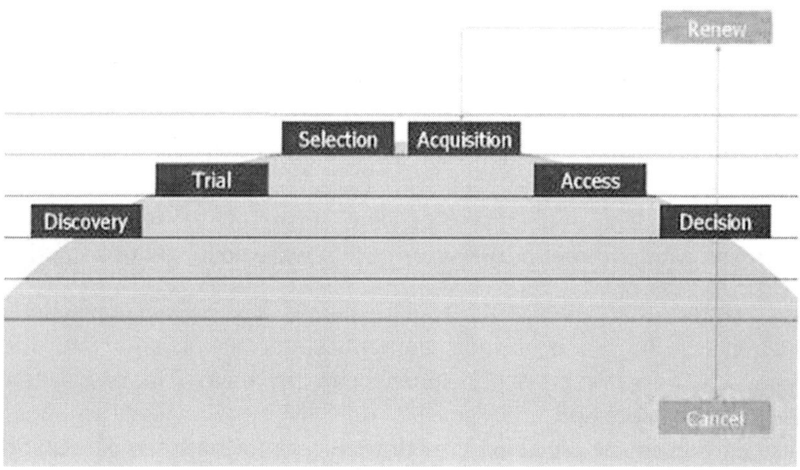

Figure 1. The e-resource life cycle

The typical stages for resources not freely available would include the following:

Discovery: The awareness of a new e-resource originates from a faculty member's request, a recommendation from a subject librarian, an advertisement, a message in a forum, or another source. The librarian then locates information about the e-resource in the ERM system's global knowledge base; information that might include, for example, the bibliographic details of an e-journal, the coverage period available, the packages that include it, and the interface or interfaces through which such packages are offered.

Trial: In many cases, the librarian will want to try out an e-resource before reaching a decision about whether to purchase it. A trial enables the librarian to offer the e-resource to some or all users - who may include patrons and librarians alike - and then base a decision on their feedback. In the trial process, the librarian activates the e-resource in the desired areas of the library environment, notifies the relevant audience, and obtains feedback.

Selection: Once the trial is over, the librarian decides whether to acquire the e-resource. A decision not to purchase the e-resource results in its deactivation in the library environment (if it had been activated as part of the trial process).

Acquisition: If the librarian decides to go forward and subscribe to the resource, then the acquisition process somewhat resembles the process used for print resources; however, an additional level of detail is required, such as license information and information about the availability of the resource to various populations of users.

Access: Access is a major issue when you are dealing with e-resources, unlike print holdings. Once a library has acquired an e-resource, the librarians want to ensure that it is well used. First, they need to make certain that users can access it easily - for example, from an A-Z list; from the OPAC, if relevant; from a metasearch tool; or via a link server. Issues such as access by remote users (for instance, proxy setup) also need to be solved. After the initial configuration of access, which might have been taken care of, at least partially, at the trial stage, librarians must deal with maintenance - including routine maintenance, problems such as the temporary unavailability of the resource, and changes in the provider's address or the manner of access.

Decision to renew or cancel: An e-resource subscription is typically valid for a defined time period. When the period ends, the librarian needs to decide whether to renew the subscription or cancel it. Unlike the initial decision at the selection phase, this decision is based on the information accumulated in the system, such as the actual usage of the resource while it was available, the reliability of the interface, and the responsiveness of the provider. Whatever the outcome of the decision - renewal or cancellation - the system needs to support it. Furthermore, even after a subscription has been cancelled, the library might have perpetual access or archiving rights to the data, another area that librarians must deal with on an ongoing basis.

The description provided here is admittedly a simplified version of the stages involved. Many e-resources today are purchased through consortia, which wield considerable purchasing power. In a consortial environment, the procedures involved in acquisition, access and decision-making are much more complex. Verde, like the other tools in the Ex Libris suite, addresses the heterogeneous needs of these consortia, particularly the differing levels of information sharing and the diverse management requirements - centralized versus localized or a combination of the two. With a collection-management tool, consortium members should ideally be able to see what other members have acquired and are considering buying and to share resources when relevant.

PARTNERING WITH THE COMMUNITY

Ex Libris has been involved in the e-resource environment for a number of years, developing products that relate to various aspects of e-resource management. These include, in particular, the use of e-resources as sources or targets of linking and as resources for metasearching; and the way in which such resources are represented to end users. Because many of the Ex Libris customers are members of consortia, Ex Libris products already address issues related to consortial handling of e-resources.

SFX®, the original link server from Ex Libris, deals with many of the complexities of managing e-resources - particularly e-journals, which pose the greatest management challenge in the electronic world. Although developed to put linking into the hands of the librarian, SFX has in many institutions become the central repository of all e-journal information. SFX tools have been developed to automatically generate e-journal A-Z lists, for example, or provide unmediated document delivery.

MetaLib®, the Ex Libris library portal, provides resource-discovery tools and metasearch (cross-database search) capabilities. The focus of MetaLib is on databases and packages of e-journals and e-books rather than on individual constituents that might be required, for example, by a linking server for accurate linking. Nevertheless, MetaLib includes tools that relate to such constituents, such as alphabetical lists of e-journals.

The adoption rate of SFX and MetaLib has proved to be quite astonishing. Three years after the first customer went live, over 670 customers in 32 countries had licensed one or both products. [2]These products have answered a significant market need and successfully address the issues involved in the access of electronic resources.

The users of ALEPH®, the integrated library system from Ex Libris, handle another aspect of e-resource management through the ALEPH OPAC. They often manually catalog e-resources or import their e-journal holdings from SFX. The imported data consists of basic information or MARC-enhanced information (via the Ex Libris MARCit! service). Furthermore, some libraries use the ALEPH acquisition module for subscribing to e-resources.

However, this set of tools does not address all aspects of e-resource management. Responding to a paper presented by the UK members of ICAU, the international consortium of ALEPH users, at the organization's 2002 meeting in Paris, Ex Libris started working with a focus group selected by Ex Libris users. Through this co-operation, Ex Libris explored the issues, defined the missing functionality, and considered ways of addressing it.

The Verde ERM system adds an important component to the Ex Libris suite of tools, primarily the management of business activities associated with the acquisition and licensing of e-resources and tools for collection development, decision making, and cost analysis. The Verde database enables libraries to maintain comprehensive information related to e-resources in one accessible location. A stand-alone system that is also readily integrated into a library environment, Verde is likely to rapidly become a focal point of the digital library.

ENTITY-RELATIONSHIP MODEL

Paying close attention to the development of the Digital Library Federation's Electronic Resource Management Initiative (DLF ERMI) committee's model, Ex Libris, with input from an international focus group, created an entity-relationship model that describes the various entities in the e-resource environment. Developed in parallel, these two models were somewhat influenced by each other, and, hence, the Verde model is actually a realization of the DLF vision (DLF, 2004). In addition, the Verde model includes two aspects that are not present in the DLF entity-relationship model: consortial support and cost analysis.

Verde is a tool for librarians who deal with e-resources. Reflecting the entity-relationship model, Verde's interface provides a sophisticated mechanism for navigating the complexities of e-resources, enabling librarians to view the various entities and their attributes and access related items. For example, a librarian can select an e-journal in one e-package and navigate to other e-packages to see the same e-journal. At each point, the librarian - if authorized - can add or modify information as necessary. In addition, the librarian can choose to follow a workflow to accomplish a specific task, such as managing a trial. In such cases, the system guides the librarian through the required steps until the workflow is complete.

Designing the interface of Verde with comprehensive navigation capabilities and workflow-oriented functions has been a great challenge. No less important is the design of the knowledge base that serves as the heart of the system. This knowledge base, which is rooted in the SFX and MetaLib knowledge bases and contains additional information, provides the backbone for the discovery functionality required for collection development and supports the appropriate Verde workflow for setting up new e-resources in the library. Although the Verde concept is that of a stand-alone system, libraries that already use SFX and MetaLib will benefit from their localized knowledge bases, which will expedite the Verde implementation process.

From the outset, Ex Libris designed Verde to support the various consortial models. Its data model permits libraries to share common data while maintaining their local or campus-specific data as necessary. Institutions can implement the model that best suits their needs, whether their administration is local, centralized, or a combination of the two.

BUILDING VERDE INTO THE LIBRARY ENVIRONMENT

Various tools already provide some of the functionality required for managing e-resources and end users' access to them. For example, end users can access an e-journal via an A-Z list, library portal, library OPAC, or link server. Verde has been constructed to integrate with such tools. An example of such integration is a Verde workflow that accommodates the activation of a specific e-package in other tools, thus enabling the user to link to the e_journals provided in the package.

Figure 2 shows how the Verde ERM system uses functionality from various components of the library environment to manage the trial, acquisition, and renewal of an e-resource.

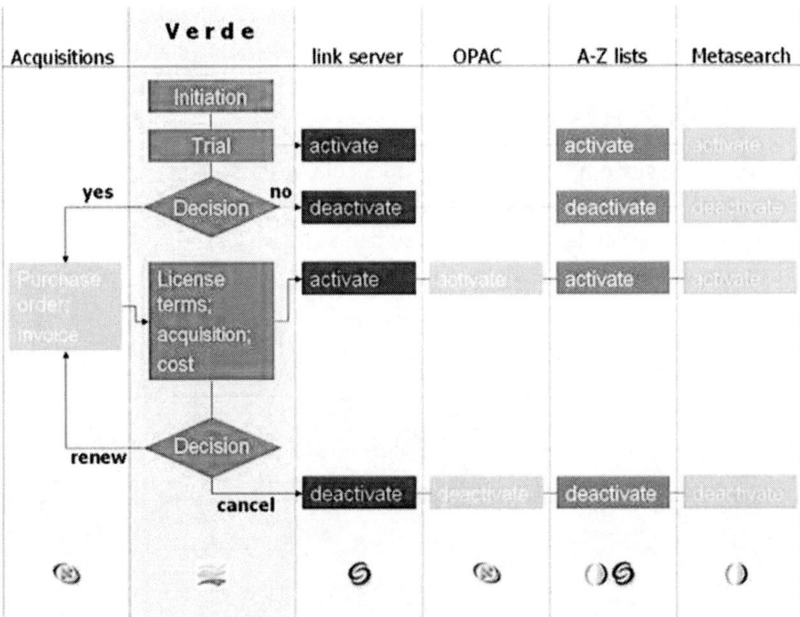

Figure 2. Verde integrated into a library environment

As Figure 2 demonstrates, the Verde system is the linchpin of e-resource management activity. Verde interacts with other systems to deal with the various stages of a resource's life cycle. For example, setting up an e-resource at the trial stage typically results in the activation of the e-resource in SFX and presumably also in an A-Z list (created by SFX or in MetaLib); when relevant, the e-resource is activated in MetaLib as well. During the acquisition stage, the new e-resource might also be presented in the OPAC. Cancellation, on the other hand, results in the deactivation of the e-resource in all the components of the library system (but perpetual access or archiving rights to the data are retained when relevant).

Even though other systems provide some of the functionality, such as acquisition, Verde logs all the information. Storage of such information in one place is crucial for providing a tool for making decisions based on cost analysis and a thorough examination of all related factors.

Verde can interact not only with Ex Libris products but also with similar products provided by other vendors. This interaction is based on industry standards, and hence implementation of Verde is the same in a non-Ex Libris environment.

VERDE ARCHITECTURE

A stand-alone system, Verde is interoperable with Ex Libris and third-party applications. These applications can communicate with Verde through either a Web service layer or SFX: they can send an OpenURL to SFX, which, in turn, provides Verde services.

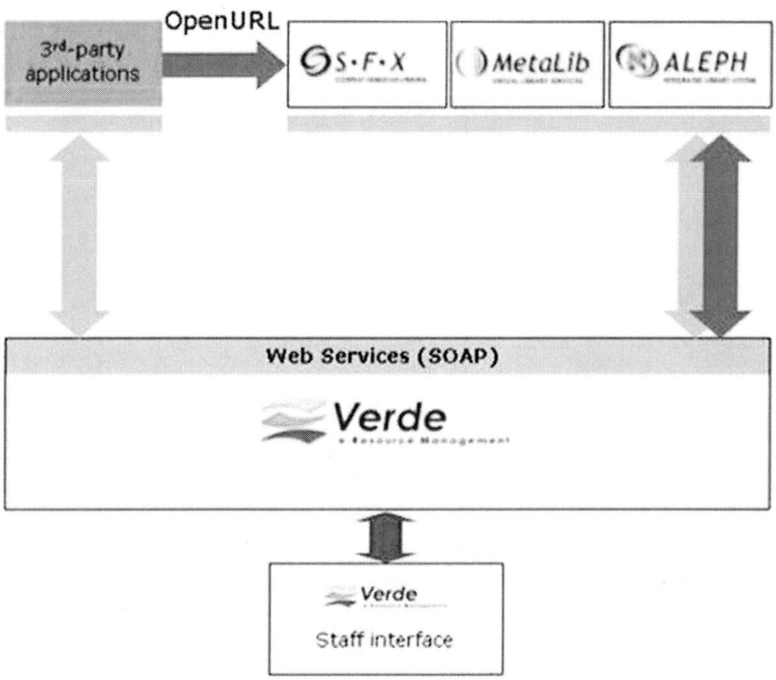

Figure 3. Verde architecture

VERDE FOR COLLECTION DEVELOPMENT

The role of the Verde ERM system as a collection-development tool cannot be underestimated. With the global e-resource offerings described in the universal e-resource knowledge base, the library's e-resource collection described in the local Verde database, and the library's expenditures expressed in the cost entity, librarians can find new e-resources of interest; take advantage of information already available in the system when adding local information; and, when deciding whether to acquire, renew, or cancel an e-resource, use existing information related to that e-resource and others.

In a consortial environment, these capabilities provide added value, enabling librarians to obtain information from other member institutions. For example, a librarian can learn about other institutions' experiences with a particular e-resource and even benefit from their agreement with the vendor.

CONCLUSIONS

Like other Ex Libris products, the Verde ERM system is a direct response to the market's needs and the outcome of a joint effort by Ex Libris, focus group participants, and development partners. The focus group and development partners have provided comprehensive information about the market and the workarounds that librarians frequently develop to deal with the complexity of e-resources. In addition, the DLF ERMI committee has made an invaluable contribution in setting the groundwork for a common understanding of the entities and workflows in the e-resource arena and for the development of standards for e-resource management (Digital Library Federation, 2004).

The timing of the Verde development process has ensured that the system complies with industry initiatives: it has been designed to adhere to the DLF ERMI model and also offer consortial support and cost information. And, of course, it is compatible with existing industry standards such as MARC, Unicode, XML, OpenURL, and SOAP. Future versions of the Verde system will support emerging standards for serials, such as ONline Information eXchange (ONIX).

There is little doubt that the industry is waiting for good tools to take the lead in the electronic resource management area. Ex Libris is confident that Verde is just such a tool.

REFERENCES

1. Anderson, Ivy, Sharon E. Farb, Adam Chandler, Timothy Jewell, Kimberly Parker, Angela Riggio, and Nathan D. M. Robertson: Report of the DLF Electronic Resource Management Initiative: Appendix A: Functional Requirements for Electronic Resource Management. Digital Library Federation, 2004.http://www.library.cornell.edu/cts/elicensestudy/dlfdeliverables/DLF-ERMI-AppendixA.pdf.
2. DLF: Electronic Resource Management Deliverables. 2004.http://www.library.cornell.edu/cts/elicensestudy/dlfdeliverables/home.htm
3. Raschke, G. and S. Weiner: 'Clarity in the Mist'. Library Journal, July 15,2004.http://www.libraryjournal.com/article/CA428134
4. WEB SITES REFERRED TO IN THE TEXT
5. Aleph®. http://www.exlibrisgroup.com/aleph.htm
6. D-Lib Magazine. http://www.dlib.org/
7. DLF - Digital Library Federation. http://www.diglib.org/dlfhomepage.htm
8. DLF ERMI - Electronic Resource Management Initiative.http://www.diglib.org/standards/dlf-erm02.htm
9. HighWire Press. http://highwire.stanford.edu/
10. MEDLINE. http://medline.cos.com
11. ONIX - ONline Information eXchange. http://www.editeur.org/onix.html
12. OUP - Oxford University Press. http://www.oup.co.uk/
13. MetaLib®. http://www.exlibrisgroup.com/metalib.htm

14. SFX®. http://www.exlibrisgroup.com/sfx.htm

15. Verde. http://www.exlibrisgroup.com/verde.htm

NOTES

[1] As this article was going to press, OUP announced that it had signed an agreement for HighWire Press to host the entire OUP journal collection, beginning in January 2005 (http://www3.oup.co.uk/jnls/press/2004/08/18/index.html)

[2] Figures as of August 2004.

Chapter 4

Information Handling in Collaborative Research

Ellen Collins, Michael Jubb

Research Information Network, GB

ABSTRACT

UK public policy makers have a growing interest in collaborative research, where academics work with public, private or third sector partners on a joint project which supports the partner's aims. This paper reports on the findings of five case studies, looking at how information is sourced, managed, used and shared within collaborative research projects. It finds that researchers within collaborative projects have similar information management issues as are known to exist within academia more broadly, but that the specific conditions which govern research collaborations mean that interventions to improve or support information management must be carefully tailored.

Keywords: knowledge economy, collaboration, information handling

All research is inherently a collaborative enterprise, building upon the expertise of colleagues, predecessors and competitors to arrive at new discoveries and understandings. In recent years, however, 'collaboration' has taken on a new meaning in UK public policy as it relates to the research sector. Increasingly, the term has come to refer to the relationships and projects developed between academic researchers and their counterparts in the public, private and third sectors. A key feature of these relationships is that they should be targeted towards a specific outcome, solving problems in order to benefit industry, society and the economy (Lambert, 2003). Collaborative research is one element of a wider spectrum of interactions between higher education institutions (HEIs) and external organisations, including consultancy, education and training, contract research and intellectual property deals: all of these have become increasingly important aspects of UK higher education funding.

The wider context of UK higher education policy has reflected and supported this drive for some time. During the 1990s, government began to encourage universities to become more proactive about encouraging the commercialisation of their research in-house and through collaborations with external organisations. A series of Higher Education Funding Council for England (HEFCE) interventions established funds to support such work. The Lambert Review and the follow-up Science and Innovation Investment Framework demonstrated the Labour government's continued interest in this area, as did the departmental reshuffle in 2007 which brought higher education and business into a single portfolio for the first time, signalling a clear agenda for academia by separating it from primary and secondary education (Lambert, 2003). The Sainsbury Review and subsequent Innovation Nation white paper focused on the business demand side, while a series of concurrent reviews looked at the skills that graduates needed to meet employers' business demands (Sainsbury, 2007). The 2011 Innovation and Research Strategy for Growth (BIS, 2011) indicates the Coalition Government's continued interest in this area, as does the Wilson Review, which reported in early 2012 (Wilson, 2012).

The long history of political support has borne fruit, in the form of a growing number of relationships between research establishments — HEIs and others — and public, private and third sector partners. The most recent data from HEFCE indicates a 4% growth in knowledge exchange between 2008–9 and 2009–10 among UK HEIs, and a growth in income from such sources by 35% from 2003–4 to 2009–10. Collaborative research income increased by just over 2% between 2008–9 and 2009–10, although changes to the survey methodology mean that this figure is not entirely reliable. The report also highlights the growing importance of non-commercial collaborators, such as government bodies or third sector charitable organisations, as partners in these enterprises. (HEFCE, 2011). A separate survey, undertaken by the Department for Business, Innovation and Skills (BIS), looks at public sector research establishments (PSREs) and their research interactions with external organisations. The most recent edition of this survey found that 25,000 new collaborative research projects began in 2007–8, and that 93% of PSREs were doing collaborative research (BIS, 2009).

So collaborations, as one possible interaction between research establishments and external bodies, have become increasingly important to universities and their finances. The question then arises: how should university librarians ensure that they engage with collaborative projects and contribute to successful research outcomes? To understand this, we must first understand the specific circumstances which govern collaborative research. Previous studies have shown that a researcher's environment and collaborators can strongly affect their information behaviour, but that they also retain strong associations and preferences from their own, early, training (RIN/NESTA, 2010; Bulger et al., 2011; Meyer et al., 2011). The cultures in business and higher education are recognised to be rather different, and the relationships can be difficult to manage (Lambert, 2003).

There are several stages of information handling for research projects, each of which offers challenges for researchers used to working in different environments. A new research project involves sourcing information that has already been published on the subject. Researchers outside academia generally have worse access to published scholarly information than those within academic paywalls, and employ different coping strategies when they cannot access the information that they need (CIBER, 2011). At this early stage, researchers will also often seek to discover and share unpublished information which may have a bearing on their research project. But researchers are often very protective of unpublished research and data: in academia, because they wish to protect future publication and funding opportunities, and in industry because of the need to protect commercial interests. Furthermore, there is some confusion and concern within industry about Freedom of Information legislation, and their vulnerability under such legislation if they share confidential data with an HEI. During the research process, new information is discovered and must be shared between different parts of the research team. Although there are standards and protocols for data collection and sharing in many disciplines, at an individual level researchers' behaviour can be idiosyncratic, and there is no guarantee that their techniques will translate into a more commercial research environment (or, indeed, that the commercial techniques will make sense in academic settings). Finally, academic researchers want to publish and share their findings, as the chief way of securing recognition and credit for their work. In many cases, this may actively undermine more commercial strategies, where there are uncertainties about the ramifications of such publication for copyright and intellectual property (CIBER, 2010). And non-commercial partners, such as public sector or charitable organisations, may want to publish findings more rapidly to reach a wide community of stakeholders, thereby diminishing the academics' chances of having an article accepted in a highly-rated journal.

The circumstances governing information handling in collaborative research are thus highly complex and nuanced. If librarians and information professionals are to engage successfully with such projects, they must understand exactly how these circumstances affect the course of individual research projects. To promote such understanding, the Research Information Network commissioned a set of case studies, which looked at five collaborative research projects to identify some common themes and issues, and some possible actions for library staff in engaging with such projects (Jordan et al., 2011). This paper considers the findings of the study.

2. METHODS

Five cases were selected, each of which included at least one university and one external partner. Overall, the cases were selected to reflect a range of criteria which we considered likely to affect information handling behaviour: these included the discipline of the collaboration, the number and types of partners involved, the stage of the collaboration — from early-stage to completed — the scale of the project and the types of external support received. Table 1 provides a brief overview of the five cases. Within each case, initial interviews with project managers helped establish the broad structure of the collaboration. Subsequent interviews with key individuals, identified in the initial stage, provided greater insight into the information management process within the collaboration. Finally, we undertook a series of workshops across the five collaborations to identify recommendations for future best practice.

It is important to note that these cases represent nothing more than a snapshot of a few research collaborations. It would be unwise to draw far-reaching conclusions on the basis of such evidence. Nonetheless, we are confident that they represent a useful insight into researcher behaviour, which can be used to support investigations and interventions by library and information professionals in universities.

Table 1. Participant cases.

	Discipline	Aim	Partners
Intelligent Care	Technology	Developing a monitoring system for older people living in residential dwellings	1 commercial partner 1 university
UK Reactics	Engineering	Exploring the potential commercial applications of novel auxetic materials (materials which expand when stretched)	6 commercial partners 3 universities
PRISMA	Health / Social care	European priorities for research and measurement in end-of-life care	11 partners, including 4 universities, with a range of public and third sector bodies
Structural Genomics Consortium (Oxford)	Structural biology	Pre-competitive mapping of protein structures to support new drug discovery	11 funders – mix of public and charitable funding and pharmaceutical companies 3 universities
Locating Communications Heritage	Arts / Digital technology	Developing a platform which enables mobile phone users to access history of communications and IT	1 university 1 third sector (museum) 2 commercial

3. THE CHALLENGES FOR INFORMATION HANDLING

The research uncovered a number of challenges for effective information handling in research collaborations. Many of these will be familiar to library and information professionals, who already encounter indifference and ineffectiveness among many of their academic researchers. Others relate more closely to the specific conditions within research collaborations.

The way that information is handled in collaborative research depends upon a number of factors. Perhaps most important among these is the primary purpose of the collaboration, and the nature of the area that it is investigating. Unsurprisingly, a collaboration which is focused upon creating a commercial product handles information differently from one which hopes to influence practice and behaviour among a user community. The effect of these differences is most readily observed around the dissemination of information. The Intelligent Care project focused upon developing new technologies which were not patentable; therefore, academic researchers on the project were very limited in their ability to publish early-stage research, as there was nothing to prevent the company's competitors using their findings to bring an alternative product to the market. UK Reactics, while also focused upon a commercial outcome, was producing research that was more likely to lead to a patent and so publication of early results was more acceptable. Although the commercial partners had 28 days to decide whether to veto a proposed publication in order to protect their intellectual property, only one publication was significantly delayed under these terms. In contrast to these two commercial projects, PRISMA sought to disseminate its findings as widely as possible, and in this case the pre-project agreement actually protected the academics, ensuring that they were able to publish findings before they were distributed among PRISMA's stakeholder community. The SGC represented a very interesting case, in that it required a high level of openness in order to protect the long-term commercial aims of the collaboration: in order to secure input from rival pharmaceutical companies, the researchers had to guarantee that no company would gain a competitive advantage by seeing the findings before any others, and the simplest way to do this was to release them to the funders, the companies and the world at large at exactly the same time. Information handling was also visibly affected by project purpose at an earlier stage in the collaboration: researchers working on Intelligent Care, for example, had to spend time investigating the market for a product rather than focusing exclusively on the academic literature.

Established cultures and norms were also important in defining information behaviour; in our five cases, there were noticeable differences between academics and their non-academic partners. For example, within academia there was a prevailing assumption of openness — researchers would share information not only with their own team but also with teams outside their own institution, even where these teams may (in some circumstances) actually be their competitors. Commercial organisations, in particular, were much more circumspect about sharing information, and tended to restrict this to internal sharing only. Aware of the greater openness within academia, they often showed reluctance to share information, rarely releasing it unless in response to a direct request, 'blinding' datasets and in some cases even generating dummy datasets to replace real, confidential, ones. Participants also observed differences in communications habits and behaviours, and the adoption of new technologies, between the academic and non-academic sectors. Academics were perceived by some of their commercial partners as having a different communications culture from the commercial world: one which was slower and more reliant on old technologies such as telephone or face-to-face conversations. In some cases, though, established norms and cultures could be very similar in the academic and non-academic organisations: the pharmaceutical companies and the academic laboratory working on the SGC project both reflected the strong conventions of this particular scientific field and used the same protocols, databases and electronic lab notebooks.

Project aims and partner cultures exert an important influence on information behaviours, but they also interact with several other characteristics of a collaborative project to further determine the way that researchers find, manage and share information. One such characteristic is the level of trust between the project partners, often based upon their previous experience of collaborating. This trust helped to mitigate some of the challenges which arose out of different expectations or experiences, and to make it easier to share information: as one researcher on the PRISMA project put it, 'it's much easier when you're trying to badger people for things if you have that level of contact [knowing each other personally]'. A commercial partner on the UK Reactics project said 'you need to reinforce with the academic what can and can't be discussed ... as you build that trust up you come into a position where you can divulge more and more to them.' These personal relationships were usually based upon previous experience of working together, or were created through a concerted effort at the start of the project through face-to-face meetings or mutually-trusted intermediaries. Previous experience of collaboration also helped partners to make allowances for the different working cultures, as they had a better understanding of drivers within organisations that were different from their own.

Alongside these informal, trusted relationships, most partnerships had more formal agreements in place which affect information management, usually agreed at the project outset. These arrangements might include non-disclosure agreements, formalised reporting and update structures and systems for monitoring project outputs. Often, these arrangements were put into place at the instigation of project funders: the Technology Strategy Board (TSB) was particularly keen to ensure information flow is managed in an effective way. In other instances, the arrangements might stem from the project requirements: SGC had strong information governance policies because of the highly competitive nature of its research and the need to ensure all partners were treated equally with no unfair advantages. The size and geographic spread of the collaboration also affected the formality of information governance arrangements: PRISMA, for example, a multi-national research collaboration with over 30 project researchers, constructed much more formal arrangements than Locating Communications Heritage, a small project with around five researchers based in London. The burden and legal cost of creating and observing a very formal information governance structure was also a consideration for Locating Communications Heritage researchers when they decided to maintain a much more informal arrangement.

Thus, the information environment in which researchers operate is affected by the collaboration's background, aims and objectives, and organisational cultures. Researchers' information activities reflect the specific circumstances of their collaboration, as outlined here, but there are some attitudes and behaviours which seem to permeate many of the collaborations that we studied.

First among these is a general indifference to — and in some cases, scepticism about — the importance of effective information handling, particularly around data management. Formal information governance policies — where they existed — were usually quite selective in what they covered. Non-disclosure agreements, intellectual property ownership and embargo procedures for scholarly publication were considered important as disagreements in these areas could potentially derail the collaboration; poor data management and curation was not seen as a threat in the same way. Furthermore, as projects were so outcome-focused, relatively little attention was paid to the long-term reusability of the data: this was not a success factor and therefore considered unimportant. The only exception to this was the SGC, where data publication was a key project aim. There was little evidence that failure to plan for and effectively manage information had a negative impact on any of the projects, but researchers did acknowledge that they might benefit from better information management — with the usual caveat that 'we shouldn't put any additional constraints or paperwork for us'.

Researcher expertise around systems and technologies for managing information, including data, was fairly limited. There was little evidence of researchers using centralised data repositories, virtual research environment, wikis or other systems to manage their project information and communications, although again the SGC was something of an exception due to the project aim of publishing data. Project support officers acknowledged that there was a lack of capacity and expertise, particularly in smaller organisations, but it was also clear that many participants had misconceptions about the repercussions of using online tools to manage information. Researchers expressed concerns about whether repositories could be hacked into, uncertainty about whether uploading information to an online system constituted 'disclosure' from the point of view of intellectual property, and concern that the company which owned the technical system might also claim ownership over any content that the research team put into it. These concerns were mostly borne out of uncertainty rather than fact: but researchers were also worried that implementing new systems might be costly and inefficient, particularly if they were overspecified.

Another important area of commonality across the projects was a limited use of library and information services, and a general perception that the benefits of using such services would be outweighed by inconvenience. As has been the case in other research projects, many researchers were not aware that the library was responsible for providing the published information that they used, and in discussing library services they were referring more to those which are accessed within the physical library space (Meyer et al., 2011). Library services were seen as too general, designed for students rather than researchers, and inexpert in providing the commercial information necessary for many projects. Furthermore, some participants saw the laborious selection and filtration of material as an essential part of the academic's work at the start of a project, helping them to become grounded within the field, and felt that asking a librarian for more effective search strategies might interfere with this process. While there was some use of specialist libraries by researchers who had previous experience of them through other projects, this did not usually spread to other project team members.

One library service which researchers did use — although, as we have said, in many cases they were not aware that they were doing so — was online access to published scholarly information. As might be expected, levels of access varied enormously between partners, and the relationships could be very complex. In some cases, a large commercial organisation might have better access than a non-research-intensive university. For this reason, along with the habitual openness in academia and the — incorrect — perception that the collaboration's confidentiality agreements protected all information exchange within the partnership, researchers often shared the full text of scholarly articles, despite copyright restrictions. Researchers also adopted coping strategies such as taking advantage of walk-in access at information-rich institutions — this, clearly, being a library service that they did value. In other cases, it was not considered necessary for all researchers to have access to the original literature, and much of the academics' time was taken up with producing literature reviews for dissemination to the wider collaboration. In these cases, participants often highlighted a need for strategies to help filter and refine the information to make it manageable and relevant.

4. THE ROLE FOR LIBRARY AND INFORMATION SERVICES

The previous section shows that researchers in collaborations have some fairly generic problems with maintaining effective information behaviours. Concerns about data management, information overload, new technologies and systems and the relevance of library services have all been found in previous studies of researchers working solely in academia, and librarians have begun to come up with some ways of addressing these concerns. However, as we have seen, the specific circumstances of collaborations can mean these problems present themselves in a more complex way, and that any attempts to intervene will need to take account of such circumstances. The scope of this study did not include extensive work with librarians and, as such, it would be inappropriate for this paper to make specific recommendations about what librarians should and should not do in working with collaborative research projects. But it is possible to highlight some important factors which might affect the success of any such interventions.

The research highlighted the importance of trust in collaborative projects, and its central role in securing input from partners. Trust was built through previous knowledge of partners, their introduction by a trusted intermediary, and evidence that they understood the drivers, behaviours and backgrounds of the various parties working within the collaboration. Once trust was established, information flowed more freely around the collaboration and researchers found it easier to work together. There was even some evidence that trusted members of a collaboration could begin to influence information behaviours: the PRISMA project manager, for example, was rigorous about following up information policies which 'dramatically changed the way that people worked'. Informal channels, rated by many researchers as more important than formal communication techniques, were heavily dependent upon trust. Any librarian or information professional seeking to work effectively with a collaborative project will need to ensure that they become a trusted part of the team. Early engagement with the project is crucial: this could be achieved by building links with knowledge transfer teams within universities, who will help facilitate many new collaborative projects.

Collaborative research projects are also noticeably outcome-driven, with a strong focus on the agreed outputs. This is accepted by the project's academic researchers, and they recognise that more traditional academic research outputs such as publications may have to assume a smaller role, if they go against the project aims. That said, many of them recognise that alternative benefits can flow from participation in collaborative research — whether that be a better understanding of industrial processes and the real-life applicability of academic research (Intelligent Care) or by participating in a prestigious collaboration (SGC). But the agreed output remains central to research activities, and systems and procedures therefore need to support these outputs. Services offered by librarians will likely have most success where they can show that they will support the aims and objectives of the project. It is clear that there are some such areas — for example, helping researchers to find and organise published information in an efficient manner, and perhaps also working with them to manage new data in a better way to improve project efficiency. But any such interventions must be proportional: the non-academic partners in particular are very wary about over-burdening themselves with unnecessary administration. Services will need to show that they can improve project outcomes and are tailored to the project's specific needs, even when this might mean that they fall short of best possible practice. Like the academic researchers who accept that they may not be able to publish, librarians will need to be a bit pragmatic in order to get their foot in the door: once they are there, they may be able to demonstrate the value of conforming fully with best practice and standards — or they may find that the standards which apply to academic research are less important when it comes to the aims and objectives of a research collaboration.

5. CONCLUSION

The five case studies showed that collaborative research does, indeed, operate upon its own terms. While strong disciplinary conventions, such as those in the scientific field explored by the SGC, can help academics and external partners work to the same standards and processes, in other cases local systems and behaviours prevail, meaning that researchers in different organisations end up working in quite different ways. This presents a challenge for librarians and information professionals in seeking to support researchers in collaborations, as they will be working with a number of different cultures, each of which has its own aims and objectives.

The aims and objectives of the collaboration as a whole are equally important in understanding how best to support information management and effective information behaviours. Collaborative research projects are particularly focused upon specified outcomes, and researchers in academia and external organisations direct their efforts towards achieving those outcomes. Any interventions by library and information professionals will need to demonstrate that they can contribute towards such outcomes, and will not produce additional burdens which might — as the researchers perceive it, anyway — distract from them.

Many of the information behaviours observed in these five cases will be familiar to library and information professionals: lack of awareness of library services, poor data management, attachment to known systems and suspicion about the time and effort required to begin using new ones. But the research also suggested that, in order to influence behaviour, external parties need to become trusted partners within the collaboration, and that a key way to do this is to demonstrate an understanding of the unique circumstances which govern its activities — its size, geographic dispersal, aims and objectives, participants' history of working together, and the relationship between formal and informal information channels. By recognising these conditions, and tailoring a service which takes account of them, librarians may be able to provide valuable support in making research collaborations more effective.

REFERENCES

1. BIS (2009). Fifth annual survey of knowledge transfer activities in public sector research establishments: Final report. London: Department for Business, Innovation and Skills. Retrieved October 6, 2012, from http://www.bis.gov.uk/assets/biscore/corporate/migratedd/publications/2/2009-5th-psre-report-v2.pdf.

2. BIS (2011). Innovation and research strategy for growth. London: Department for Business, Innovation and Skills. Retrieved October 6, 2012, from http://www.bis.gov.uk/assets/biscore/innovation/docs/i/11-1387-innovation-and-research-strategyfor-growth.pdf .

3. Bulger, M. E., Meyer, E. T., de la Flor, G., Terras, M., Wyatt, S., Jirotka, M., Eccles, K., & Madsen, C. (2011). Reinventing research? Information practices in the humanities. London: Research Information Network. Retrieved October 6, 2012, from http://

www.rin.ac.uk/our-work/using-and-accessing-information-resources/informationuse-case-studies-humanities.

4. CIBER (2010). Research support services in UK universities: What services do researchers need and use? London: Research Information Network. Retrieved October 6, 2012, from http://www.rin.ac.uk/our-work/using-and-accessing-information-resources/ research-support-services-what-services-do-resear.

5. CIBER (2011). Access to scholarly content: gaps and barriers. London: Research Information Network. Retrieved October 6, 2012, from http://www.rin.ac.uk/ node/1172.

6. HEFCE (2011). Higher education – Business and community interaction survey 2009–10. London: Higher Education Funding Council for England. Retrieved October 6, 2012, from http://www.hefce.ac.uk/media/hefce1/pubs/hefce/2011/1125/11_25.pdf.

7. Jordan, E., Hunter, A., Seale, B., Thomas, A., & Levitt, R. (2011) Information handling in collaborative research: an exploration of five case studies. London: Research Information Network. Retrieved October 6, 2012, from http://www.rin.ac.uk/system/files/ attachments/Collaborative_research_report.pdf.

8. Lambert, R. (2003). Lambert review of business-university collaboration: Final report. London: HMSO. Retrieved October 6, 2012, from http://www.hm-treasury.gov. uk/d/lambert_review_final_450.pdf.

9. Meyer, E. T., Bulger, M., Kyriakidou-Zacharoudiou, A., Power, L., Williams, P., Venters, W., Terras, M., & Wyatt, S. (2011). Collaborative yet independent: Information practices in the physical sciences. Institute of Physics, IOP Publishing. Retrieved October 6, 2012, from http://www.iop.org/publications/iop/2012/file_53558.pdf.

10. RIN/NESTA (2010). Open to All? Case studies of openness in research. London: Research Information Network. Retrieved October 6, 2012, from http://www.rin.ac.uk/ system/files/attachments/NESTA-RIN_Open_Science_V01_0.pdf.

11. Sainsbury, D. (2007). The race to the top: A review of government's science and innovation policies. London, HMSO. Retrieved October 6, 2012, from http://www.hm-treasury.gov.uk/d/sainsbury_review051007.pdf.

12. Wilson, T. (2012). Review of business-university collaboration: The Wilson review. London: Department for Business, Innovation and Skills. Retrieved October 6, 2012, from http://webarchive.nationalarchives.gov.uk/20120305154225/http://www. wilsonreview.co.uk/wilson-review/wilson-review.pdf.

Chapter 5

Knowledge Creation and Transfer Amongst Post-Graduate Students

Kreeson Naicker[1], Krishna K. Govender[1,2], Karunagaran Naidoo[1]

[1]School of Management, Information Technology & Governance, University of KwaZulu-Natal, South Africa

[2]Regenesys Business School, Johannesburg, South Africa

ABSTRACT

Background: The skill shortages, hyper-competitive economic environments and untapped economies have created a great deal of focus on knowledge. Thus, continuously creating and transferring knowledge is critical for every organisation.

Objectives: This article reports on an exploratory study undertaken to ascertain how knowledge is created and transferred amongst post-graduate (PG) students, using the knowledge (socialisation, externalisation, combination, internalisation [SECI]) spiral model.

Method: After reviewing relevant literature, a personally administered standardised questionnaire was used to collect data from a convenience sample of PG students in the School of Management, IT and Governance at the University of KwaZulu-Natal, South Africa. The data was analysed to determine if it fit the model based on the four modes of knowledge conversion.

Results: Although the School of Management, IT and Governance has mechanisms in place to facilitate knowledge creation and transfer, it nevertheless tends to focus on the four modes of knowledge conversion to varying degrees.

Conclusion: The study confirmed that PG students utilise the 'socialisation' and 'externalisation' modes of knowledge conversion comprehensively; 'internalisation' plays a significant role in their knowledge creation and transfer activities and whilst 'combination' is utilised to a lesser extent, it still plays a role in PG students' knowledge creation and transfer activities. PG students also have 'space' that allows them to bring hunches, thoughts, notions, intuition or tacit knowledge into reality. Trust and dedication are common amongst PG students. With socialisation and externalisation so high, PG students are aware of each other's capabilities and competencies, and trust each other enough to share knowledge.

INTRODUCTION

Information on demand is a powerful aspect of the mode in which academic organisations, teams and individuals operate. There is a constant need for academic organisations, teams and individuals to accelerate the communication of information and knowledge to each other and organisations outside the academic sphere (Nelson 2005). An individual is often afraid to share their knowledge and they have a tendency to guard their knowledge and selectively release it. This tendency is often cited as a core problem when working in a team and the cause of poor collaboration between team members (Gilmour 2003). In order to leverage on innovation as one of the most important sources of competitiveness and success, academics need to have access to and mobilise their knowledge resources (Voelpel, Von Pierer & Streb 2006).

Knowledge is generally separated into two types: explicit and tacit knowledge. Hautala (2011:605) explains that 'academic knowledge aims at creating or exploring the new' tacit and explicit elements, which together with theory and practice are used to form knowledge. Tacit knowledge is a personal, contextual and practical entity that is difficult to communicate (Polanyi 1966). For example, balancing a bicycle to ride it requires tacit knowledge that is not easy to explain to someone who has never actually ridden one. Similarly, understanding and conducting a research project as part of an academic field, research group and society also includes tacit knowledge (Hautala 2011:605).

Davenport and Prusak (1998:81) argue that codifying tacit knowledge is difficult, but that 'its substantial value makes it worth the effort'.

Notwithstanding some of the challenges cited by the scholars and practitioners referenced above, the decision to focus only on tacit knowledge instead of knowledge in general can be justified by reasons that include, but are not limited to, its substantial volume and value to the organisation (Suppiah & Sandhu 2011). The ability to create, store, disseminate and utilise knowledge and expertise has become a primary way for organisations, teams and individuals to compete (Hayashi 2004). Amassing and synthesising specialised knowledge from multiple sources is a pivotal factor in resolving the technical and operational uncertainties that impede Nonaka and Takeuchi's (1995) knowledge (SECI) spiral theory.

In light of the above, this article focuses on determining which of the four modes of knowledge conversion assist post-graduate (PG) students to access information on demand.

It is widely accepted that there is difficulty in disseminating tacit knowledge, since tacit knowledge is personal and understandable by the possessor (Alavi & Leidner 2001; Nonaka & Takeuchi 1995; Polanyi 1966). Thus, this study also attempts to establish, through an assessment of the knowledge spiral, if extracting tacit knowledge or engaging with explicit knowledge is essential for a paradigm shift when transcending from the old self to the new self.

LITERATURE REVIEW

The genesis of the knowledge (SECI) spiral process

Knowledge is sought and shared in a global arena, whether at a corporate or academic level (Hautala 2011). McKenzie and Van Winkelen (2011) argue that:

sound decisions rely on having the right knowledge in the right place at the right time, to be able to act effectively. 'Right' knowledge may be different for every decision – some decisions require only surface knowledge, some require more investigation and an evidence base, some use tacit expertise, and others creative insight, intuition and judgment. (p. 403)

Rai (2011) believes that despite the subtle differences between the various knowledge definitions, scholars agree that effective and efficient knowledge management is central to organisational performance and success. In order to assess the capacity of an organisational system to generate new knowledge, the first step is to define knowledge and then how to determine if it is 'new' knowledge, as the aforementioned has a justified belief that increases an entity's capacity for effective action (Arling & Chun 2011).

The conversion of tacit knowledge into explicit knowledge helps knowledge to be crystallised and shared by others, which becomes the basis for the creation of new knowledge. The successful conversion of tacit knowledge into explicit knowledge depends on the sequential use of metaphor, analogy and models (Rai 2011). The materialisation of new knowledge always begins with the individual. A resourceful individual may become conscious of a position that has not been developed, which may lead to the growth or advancement of a product, service or theory. An individual's personal knowledge is transformed into organisational knowledge that is valuable to the company. Making personal knowledge available to others is the central activity of the knowledge-creating company. Nonaka and Konno (1998:26) argue that 'it takes place continuously and at all levels of the organisation', in that one's personal knowledge is transformed into organisational knowledge through the interactions between tacit knowledge and explicit knowledge. Within several loops of interaction where community members share their experiences, ideals and ideas, new knowledge – individual as well as collective knowledge – emerges through this process (Renzl 2006). Oguz and Sengün (2011:446) argue that Nonaka and Takeuchi (1995) popularised tacit knowledge in the management literature. Using the example of the bread master, they promote the link between tacit and explicit knowledge. Their work legitimised the tacit-explicit dichotomy by viewing the two as separate spheres of knowledge. Even though they cite and use Polanyi (1966) approvingly, the ontological dimension of knowing remained inconspicuous. In their view, knowledge creation is the result of an interactive spiral between tacit and explicit knowledge. This rendition has been widely accepted in most of the following literature and has created the tendency to see tacit and explicit knowledge as substitutes (e.g. Nonaka & Takeuchi 1998). According to Nonaka and Takeuchi, there are four ways of converting between tacit and explicit knowledge; these are: socialisation (tacit to tacit), externalisation (tacit to explicit), combination (explicit to explicit) and internalisation (explicit to tacit).

Grant (2007) views tacit knowledge as the ability or skill of an individual to do something or to resolve a problem that is based, in part, on one's own experiences and learning, and probably not all of this knowledge can be shared between individuals. Capturing tacit knowledge is seen as the challenge to organisations that want to spread knowledge throughout the organisation or spur greater innovation. It is treated as a reserve deposited deep within the ground that needs to be detected and then pumped out (Mooradian 2005).

The aforementioned development led Nonaka and Takeuchi (1995), the pioneers of the knowledge spiral model, to believe that tacit and explicit are not totally separate but mutually complementary entities; this led them to further develop the notion that tacit and explicit knowledge interact with and interchange into each other in the creative activities of human beings. The interaction between tacit and explicit knowledge is known as knowledge conversion, which consists of four modes. Girard (2006) believes that the four modes of knowledge conversion need to operate in sync. The four modes of knowledge conversion consist of socialisation, externalisation, combination and internalisation, which are shown in Figure 1. It is important to stress that the success of each mode of knowledge conversion will depend on the leadership and culture of the organisation or team. After all, managing knowledge is all about creating a culture that will institutionalise trust and facilitate knowledge creation, transfer and storage (Kermally 2002). Socialisation helps to move knowledge in tacit form between individuals (in this instance post-graduate students), externalisation is the application of tacit insights on an outside entity, combination represents the act of synthesising explicit pieces of knowledge and finally internalisation is the process through which one increases one's knowledge by learning from external events (Desouza & Awazu 2006).

Tacit knowledge *to* Explicit knowledge

Tacit knowledge	**Socialisation**	**Externalisation**
from		
Explicit From knowledge	**Internalisation**	**Combination**

Source: Adapted from Nonaka, I. & Takeuchi, H., 1995, *The knowledge creating company*, Oxford University Press, New York

FIGURE 1: Four modes of knowledge conversion.

Knowledge (SECI) conversion refers to the four modes of knowledge creation identified by Nonaka and Takeuchi (1995): socialisation that transfers tacit knowledge, externalisation that converts tacit knowledge into explicit knowledge, combination that integrates explicit knowledge and internalisation that embodies new tacit knowledge (Choo & Drummond de Alvarenga Neto 2010). The knowledge conversion (SECI) process (Figure 2) is deemed to be the blueprint for the knowledge spiral model (Figure 2), where the interaction or dialogue between the modes plays an integral role in knowledge creation and transfer.

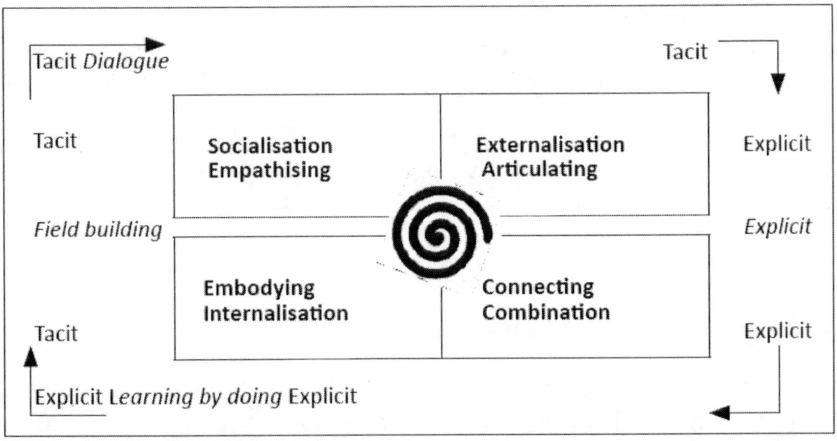

Source: Adapted from Nonaka, I. & Takeuchi, H., 1995, *The knowledge creating company*, Oxford University Press, New York

FIGURE 2: The knowledge spiral.

Arling and Chun (2011) and Perez-Araos *et al.* (2007), after an examination of Alavi and Leidner's (2001) and Nonaka and Takeuchi's (1995) research respectively, explain that socialisation is the process of converting one individual's tacit knowledge to another individual's tacit knowledge through interpersonal interaction. Tacit knowledge to tacit knowledge (socialisation) is a process of sharing experiences in a direct face-to-face approach to create tacit knowledge, often done through shared mental models, technical skills, observation, imitation and practice. Externalisation is the process of converting tacit knowledge to explicit knowledge. Tacit knowledge to explicit knowledge (externalisation) is a knowledge creation process where a part of tacit knowledge is articulated and turned somehow into explicit form, through analogies, concepts, hypothesis, models, reports, and so on. Combination is the process of creating new explicit knowledge by reconfiguring, re-categorising and re-conceptualising existing explicit knowledge. Explicit knowledge to explicit knowledge (combination) is a process of combining different bodies of explicit knowledge and internalisation is the process of converting explicit knowledge to tacit knowledge. Explicit knowledge to tacit knowledge (internalisation) is a process of embodying explicit knowledge into tacit knowledge by experiencing knowledge through the explicit source (learning-by-doing approach).

Nonaka and Takeuchi's (1995) knowledge (SECI) spiral model is reliant on the interaction between the micro and macro environments, and changes occur at both the micro and the macro level; an individual (micro) influences and is influenced by the environment (macro) with which they interact (Little, Quintas & Ray 2002).

Against the brief literature review, this study explored knowledge creation and transfer amongst PG students at a research university.

RESEARCH METHODOLOGY

An exploratory research design was employed based on Saunders *et al.*'s (2007:133) affirmation that 'an exploratory study is a valuable means of finding out what is happening; to seek new insights; to ask questions and to assess phenomena in a new light'. Furthermore, Sekaran (2003:119) argues that 'exploratory studies are undertaken to better comprehend the nature of the problem since few studies might have been conducted in that area'.

This study was quantitative in nature, since:

quantitative methods require the use of standardized measures so that varying perspectives and experiences of people can be fitted into a limited number of predetermined response categories to which numbers are assigned. (Patton 2002:14)

Furthermore: the advantage of a quantitative approach is that it's possible to measure the reactions of many people to a limited set of questions, thus facilitating comparison and statistical aggregation of the data. This gives a broad, generalizable set of findings presented succinctly and parsimoniously. (Patton 2002:14)

A questionnaire was used, which comprised a five-point Likert scale that ranged from 1–5 (with 5 = strongly agree and 1 = strongly disagree). The responses were converted to frequencies and percentages, and the results were related to Nonaka and Takeuchi's (1995) knowledge (SECI) spiral model and interpreted accordingly.

As a first step in deciding on the sample, all PG students in the School of Management, Information Technology & Governance, at the University of KwaZulu-Natal constituted the target population. The reason for choosing this group of 'knowledge creators' was a matter of convenience, since a sample from this population would have been easily accessible, since they comprised individuals who were known to the primary researcher. According to McBurney and White (2007) the sampling frame is a population as it is defined for the purposes of selecting subjects for a study; in the context of this study, the *sampling frame* is all accessible post-graduate students in the School of Management, Information Technology & Governance. The *sample size*, which required making some judgement about the number of participants needed for the study (Devlin 2006), was determined using a non-probability sampling technique, namely convenience sampling. The population of post-graduate students comprises 204 honours students and 132 master's and PhD students. The total number of post graduate students (336) was used in the calculation of the sample size. Using a confidence level of 95% with a confidence interval of 12 results in a sample size of 56 required whilst a confidence level of 95% with a confidence interval of 10 results in a sample size of 75. It was therefore decided to use a sample size within the range of 56 and 75. Approximately 60 students were targeted. The researcher collected 70 completed questionnaires.

Sekaran (2003) explains that for personally administered questionnaires:

when a survey is confined to a local area, and the organisation is willing and able to assemble groups of employees to respond to the questionnaires at the workplace, a good way to collect data is to personally administer the questionnaires. The main advantage of this is that the researcher or a member of the research team can collect all the completed responses within a short period of time. Any doubts that the respondents might have or any question could be clarified on the spot. (p. 236)

Thus, the questionnaires were personally administered to the PG students.

The reliability of a measure, which indicates the 'extent to which it is without bias (error free) and hence ensures consistent measurement across the various items in the instrument' (Sekaran 2003:203), is reliant on the stability and consistency with which the instrument measures the concept, model or theory. The internal consistency of the measure was maintained as all the sections of the questionnaire focused on measuring if the adaptation of Nonaka and Takeuchi's (1995) knowledge (SECI) spiral model has an influence on the way knowledge is created and transferred by PG students.

Sekaran (2003) proposes that validity is concerned with the authenticity of the cause-and-effect relationships (internal validity), and their generalisability to the external environment (external validity). Although there are several ways of testing validity, this study focused on content validity which is the best test of validity for a questionnaire. Content validity ensures that the measure includes an adequate and representative set of items that tap the concept, theory or model. The questions based on Nonaka and Takeuchi's (1995) knowledge spiral is an adequate and representative set of items that tap into the way knowledge is created and transferred by PG students. The questions are dimensions and elements of Nonaka and Takeuchi's knowledge (SECI) spiral model. For completeness, see the questionnaire in Annexure 1.

Empirical findings

The majority (93%) of respondents were honours students. The remaining 7% were master's students and PhD students. A total of 380 questionnaires were administered.

The Likert scale that was used in the questionnaire had five categories. For the purpose of analysis, all 'strongly agree' and 'agree' responses were collapsed into one category, namely 'agree' because it can be seen from the results that even if 'neutral' is combined with 'disagree' and 'strongly disagree', the results would not be changed.

The use of socialisation

The first mode of the knowledge conversion process, *socialisation*, is determined by seven statements. Table 1 reflects the respondents' agreement or disagreement with each statement.

It is evident from the information in Table 1, that the overwhelming majority of respondents agreed with all seven statements, implying that the 'socialisation' mode was effective in knowledge creation and transfer amongst the PG students.

TABLE 1: Respondents' use of socialization

Response	% Who agreed
'I will actively share my experience with others'	82.9
'In my academic team, my teammates and I will share life or work experience with each other'	84.3
'During group discussion, I try to find out others' opinions, thoughts and other information'	94.3
'During discussion, I will bring out some concepts, thoughts or ideas'	92.9
'I often encourage others to express their thoughts'	87.1
'Before group discussion, I will collect necessary information and show it to the group'	68.6
'I like to get to know the people with whom I will work before going into a project together'	75.7

Source: Developed by researchers from the research data

Use of combination

The third mode of the knowledge conversion process, *combination*, is determined by six statements in the questionnaire.

TABLE 2: Respondents' use of combination.

Response	% Who agreed
'During the discussion, I tend to organise ideas and make conclusions to facilitate the discussion'	74.3
'When coming across problems, I tend to use my experience to help solving problems'	84.3
'After every event, I have the habit of organising and making a summary of what happened'	58.6
'During discussion, I will organise everyone's thoughts in my mind'	61.4
'I like to collect new information and make a connection of new and old knowledge to work up new concepts'	77.1
'I like to organise ambiguous concepts into structure'	63.33

Source: Developed by researchers from the research data

It is evident from Table 2, that since a two-thirds majority was not achieved for some statements pertaining to 'combination', the 'combination' mode is not predominantly used by the PG students. This confirms that transforming tacit knowledge to explicit knowledge is not a significant activity in the knowledge creation and transfer process.

The use of internalisation

Five statements (Table 3) served to ascertain information about the respondents' '*internalisation*' of information. From Table 3, it is evident that for the majority (four-fifths) of the statements that comprised 'internalisation,' the sample was in agreement, which implies that this mode is predominantly being used by the PG students in the School of Management, Information Technology & Governance, at the University of KwaZulu-Natal.

TABLE 3: Use of internalisation

Response	% Who agreed
'After hearing a new idea or concept, I tend to compare it with my experience to help me comprehend the meaning'	78.6
'I understand others' thoughts better by repeating what they said and asking them "Is this what you mean?"'	78.6
'I will tell others what I think to make sure my understanding is the same as theirs'	84.3
'When I have finished saying something, I will ask the other person if it is necessary to repeat to make sure they understand exactly what I mean'.	60.0
When communicating with others, I will give others time to think about we discussed	77.1

Source: Developed by researchers from the research data

The use of externalisation

The second mode of the knowledge conversion process, externalisation, was determined by seven statements.

In order to ascertain whether the mode of 'externalisation' is predominantly used by the PG students, at least a two-thirds majority

(66%) agreement (Table 4) with each of the seven statements is required. The study revealed that there was overwhelming agreement for each individual statement. Therefore, the 'externalisation' mode is effective and being frequently used by the PG students. This shows that PG students are able to articulate their tacit knowledge to explicit knowledge effectively, or that there are academic mechanisms in place to help post-graduate students transform their tacit knowledge into explicit knowledge.

TABLE 4: Respondents' use of externalisation.

Response	% Who agreed
'When others can't understand me, I am usually able to give them examples to help explaining'	95.7
'Most of the time, I can transcribe some of the unorganised thoughts into concrete ideas'	68.6
'I can describe academic or technical terms with conversational language to help communicate in a group'	82.9
'I tend to use analogy when expressing abstract concepts'	84.3
'When I try to express abstract concepts, I tend to explain with examples'	94.3
'I will help others to clearly express what they have in mind by encouraging them to continue what they are saying'	87.1
'When others cannot express themselves clearly, I usually help them clarify their points'	75.7

Source: Developed by researchers from the research data

RECOMMENDATIONS

More research should be carried out on how knowledge is created and transferred by PG and undergraduate students on a larger scale, since students are the individuals that directly deal with the generation, transfer and storage of knowledge in academic institutes. Both qualitative and quantitative research should be carried out in this area using a larger sample size, at least 1000 students, depending on the student population of the university. Future research in addition should concentrate on the conditions that drive the knowledge spiral process, the five-phase model of the organisational knowledge-creation process (in an academic context) and the factors that affect the knowledge spiral process and the individual modes of knowledge. Moreover, the researcher strongly believes that a study should be done to compare new joiners in corporate university graduate programs, post-graduate students and undergraduate students. A study of the influence of trust, communication, culture, organisational networks, technical infrastructure and other influences such as investment should be carried out, to determine their effects on knowledge creation and transfer in academic institutes. Future studies should also compare knowledge creation, transfer and storage in South African academic institutions with that of other American, European and Asian academic institutions; it should be country specific. This will help to gauge if there are any similarities or differences between academic institutions other than political, social and economic differences. This would also enable future researchers to observe the cultural differences between students or academic institutions.

Future studies on knowledge conversion activities in academic institutes should focus on the years of existence, number of students, faculty, schools and the academic year of the student. Furthermore, future research should determine if students in specific years of study and who have certain access to academic knowledge mechanisms are more adept at creating, transferring and storing knowledge. Finally, as the culture of the university could not be established with a single school or a few disciplines, the author strongly believes that multiple colleges should be surveyed to determine if a knowledge creating and transfer culture exists.

CONCLUSION

The aim of this study was to assess how knowledge is created and transferred amongst PG students, using Nonaka and Takeuchi's (1995) knowledge (SECI) spiral model. The study confirmed that PG students utilise the 'socialisation' and 'externalisation' modes of knowledge conversion comprehensively; 'internalisation' plays a significant role in their knowledge creation and transfer activities too, whilst 'combination' is less utilised but still plays a role in their knowledge creation and transfer activities. PG students have space that allows them to bring hunches, thoughts, notions, intuition or tacit knowledge into reality. Trust and dedication are common amongst post-graduate students. With the use of 'socialisation' and 'externalisation' so high, PG students are aware of each others' capabilities and competencies, and trust each other enough to share the knowledge.

As 'socialisation' and 'externalisation' received the majority of the positive responses, the activities associated with these modes will illustrate the trusted source of information and knowledge. Therefore, brainstorming, informal meetings, discussions, dialogues, observation, mentoring, learning groups, as well as meetings, building hypotheses and models, pictures for communication, after-action reviews, workshops, master classes, assignment databases, best practice exchange, diagrams, illustrations, sketches, metaphors and analogies are the trusted sources of information and knowledge for post-graduate students at the School of Management, Information Technology & Governance, at the University of KwaZulu-Natal.

COMPETING INTERESTS

There are no financial or non-financial competing interests inherent in the publication of this article, other than the fact that it was extracted from the mini-dissertation of a master's student, whose study was co-supervised by the co-authors.

AUTHORS' CONTRIBUTIONS

K. Naicker (University of KwaZulu-Natal) is the author of the mini-dissertation from which this article was extracted. However, the article was primarily written by the research supervisor K.K.G. (University of KwaZulu-Natal; Regenesys Business School) and formatted and edited by K. Naidoo (University of KwaZulu-Natal), the co-supervisor.

REFERENCES

1. Alavi, M. & Leidner, D.E., 2001, 'Review: Knowledge management and knowledge management systems: Conceptual foundations and research issues', *MIS Quarterly* 25(1), 107–136.

2. Arling, P.A. & Chun, M.W.S., 2011, 'Facilitating new knowledge creation and obtaining KM maturity', *Journal of Knowledge Management* 15(2), 231–250.

3. Choo, C.W. & Drummond de Alvarenga Neto, R.C., 2010, 'Beyond the ba: Managing enabling contexts in knowledge organisations', *Journal of Knowledge Management* 14(4), 592–610.

4. Davenport, T.H. & Prusak, L., 1998, *Working knowledge: How organisations manage what they know*, Harvard Business School Press, Boston.

5. Desouza, K.C. & Awazu, Y., 2006, 'Knowledge management at SME's: Five peculiarities', *Journal of Knowledge Management*10(1), 32–43.

6. Devlin, A.S., 2006, *Research methods: Planning, conducting and presenting research*, Thomson Wadsworth, California.

7. Gilmour, D., 2003, 'How to fix knowledge management', *Harvard Business Review* 16, 31–38.

8. Girard, J.P., 2006, 'Where is the knowledge we have lost in managers?', *Journal of Knowledge Management* 10(6), 22–38.

9. Grant. K.A., 2007, 'Tacit knowledge revisited – We can still learn from Polanyi', *The Electronic Journal of Knowledge Management* 592, 173–180.

10. Hautala, J., 2011, 'Cognitive proximity in international research groups', *Journal of Knowledge Management* 15(4), 601–624.

11. Hayashi. A.M., 2004, 'Building better teams', *Sloan Management Review*, Winter, 179–186.

12. Kermally, S., 2002, *Effective knowledge management*, John Wiley & Sons, Ltd., West Sussex.

13. Little, S. Quintas, P. & Ray, T., 2002, *Managing knowledge: An essential reader*, Sage Publications Ltd, London.

14. McBurney, D.H. & White, T.L., 2007, *Research methods*, Thomson Wadsworth, California.

15. McKenzie, J. & Van Winkelen, C., 2011, 'Developing organizational decision-making capability – A knowledge manager's guide', *Journal of Knowledge Management* 15(3), 403–421.

16. Mooradian, N., 2005, 'Tacit knowledge: Philosophic roots and role in KM', *Journal of Knowledge Management* 9(6), 104–113.

17. Nelson, R.R., 2005, 'Economic development from the perspective of evolutionary economic theory', *Oxford Development Studies* 36(1), 9–21.

18. Nonaka, I. & Konno, N., 1998, 'The concept of "ba": Building a foundation for knowledge creation', *California Management Review* 40, 40–54.

19. Nonaka, I. & Takeuchi, H., 1995, *The knowledge creating company*, Oxford University Press, New York.

20. Oguz, F. & Sengün, A.E., 2011. 'Mystery of the unknown: Revisiting tacit knowledge in the organizational literature', *Journal of Knowledge Management* 15(3), 445–461.

21. Patton, M.Q., 2002, *Qualitative research and evaluation methods*, Sage Publications, Inc., California.

22. Perez-Araos, A., Barber, K.D., Munive-Hernandez, J.E. & Eldridge, S., 2007, 'Designing a knowledge management tool to support knowledge sharing networks', *Journal of Manufacturing Technology Management* 18(2), 153–168.

23. Rai, R.K., 2011, 'Knowledge management and organizational culture: A theoretical integrative framework,' *Journal of Knowledge Management* 15(5), 779–801.

24. Renzl, B., 2006, 'Trust in management knowledge sharing: The mediating effects of fear and knowledge documentation',*Omega* 36(2), 206–230.

25. Saunders, M.N.K., Saunders, M., Lewis, P. & Thornhill, A., 2011, *Research methods for business*, 5th edn., Pearson, India.

26. Sekaran, U., 2003, *Research methods for business: A skill building approach*, John Wiley & Sons, Inc., New Jersey.

27. Suppiah, V. & Sandhu, M.S., 2011, 'Organisational culture's influence on tacit knowledge-sharing behaviour', *Journal of Knowledge Management* 15(3), 462–477.

28. Voelpel, S.C., Von Pierer, H. & Streb, C.K., 2006, 'Mobilizing organizations for innovation and value creation: An integrated model of the mobile company', *Journal of Knowledge Management* 10(6), 5–21.

Chapter 6

Establishing a Research Information System as Part of an Integrated Approach to Information Management: Best Practice at the Karlsruhe Institute of Technology (KIT)

Frank Scholze, Jan Maier

Karlsruhe Institute of Technology (KIT), KIT Library, Karlsruhe, Germany

Avedas, Karlsruhe, Germany

ABSTRACT

The Karlsruhe Institute of Technology (KIT) is one of the largest research and higher education organisations in the world focusing on engineering and natural sciences. At present KIT, under the chairmanship of its executive board, is installing an extensive current research information system (CRIS) covering all institutes and facilities of the organisation. The assumption underlying the project is that a consistent overview of research performance has become fundamental for the international competitiveness of research institutions and is increasingly important for strategic decisions at the executive level. Ultimately, it also leads to better data and control in rankings at higher education assessments. The new research information system systematically maps all of KIT's processes and instruments to obtain, connect, present and utilise the research metadata of active researchers. This reduces the documentation workload for researchers, for the executive level and central units such as the library, and at the same time allows for and facilitates an overall view and the aggregation and visualisation of research metadata. Our vision is to build a federally structured network of systems that gathers information on KIT's publications, research competence, research projects, patents and technological offers by retrieving data from external and internal sources as well as directly from the researchers. The network facilitates linking and aggregating of data and provides unique identifiers for individual researchers and organizational units. With its consistent data model the research information system also fosters the organisational development of KIT, which was formed in 2009 by the merger of a university and a national research centre. The researchers and their activities are at the core of the research information system. The system substantially reduces their administrative burden in documenting project information and publications. Automatic data import from external online sources and repositories has the big advantage that publication data are acquired and validated only once and can be used subsequently for manifold purposes such as websites, CVs, publication lists and reference management software. Due to the complexity of workflows, internal and external sources and the organisational infrastructure that is required for such an implementation, KIT has chosen a commercial software solution (by AVEDAS AG) as the basic installation, and in cooperation with AVEDAS will expand it and adjust it to a tailor-made system that covers the entire KIT publication and research life cycle. The CRIS at KIT is part of a larger integrated and service-oriented approach to information management (Karlsruhe Integrated Information Management — KIM).

Keywords: research, information system, information management

THE KARLSRUHE INSTITUTE OF TECHNOLOGY (KIT)

The Karlsruhe Institute of Technology (KIT) is a German academic research and education institution resulting from a merger between the University (Universität Karlsruhe) and the Research Centre (Forschungszentrum) Karlsruhe. The university, also known as Fridericiana, was founded in 1825. In 2009 it merged with the former national nuclear research centre founded in 1956 as the Kernforschungszentrum Karlsruhe (FZK). KIT is a unique initiative as it combines the mission of a University funded by the State of Baden-Württemberg with the mission of a national research centre funded by the Federal Government. A special law had to be passed to make this happen. From a historical perspective, however, it was a logical step, as both institutions look back on a long tradition of research and education with many parallels in terms of content and organisation. KIT is one of nine German Excellence Universities and positions itself as an institution with excellent research and teaching in natural and engineering sciences on an international scale, and with scientific excellence as a global cutting-edge university in research, teaching and innovation. As a member of the Helmholtz Association, the largest science organisation in Germany, KIT makes major contributions to top national and international research.

KARLSRUHE INTEGRATED INFORMATION MANAGEMENT (KIM)

The concept of 'Karlsruhe Integrated Information Management (KIM)' pursues the goal of increasing excellence in research and teaching at the Karlsruhe Institute of Technology (Hartenstein et al., 2007). Therefore, it strives for a continuous and sound integration of relevant legacy systems and data as well as for increasing the accessibility and transparency of related business processes. The KIM project's technical realisation is based on a service- oriented aAchitecture (SOA) model which was termed 'KIM integrated Service Oriented Architecture (KIM iSOA)' (Freudenstein et al., 2006).

KIM consists of a series of associated sub-projects (Figure 1). Among them are 'Management of courses, examinations and study assistance (KIM-CM)' and 'Identity management (KIM-IDM)'. The results of these projects have led to services which are accessible through the portals 'studium.kit.edu' and 'intra.kit.edu'. Furthermore, provisioning all KIT employees and students on the basis of their associated KIT accounts not only facilitates innovative IT services, but contributes substantially to giving people in the university and in the large-scale research sector a sense of solidarity.

In order to co-ordinate information provision and processing, and to ensure that organisational and cultural change associated with the KIM projects is managed properly, co-ordination structures have been established at KIT. Among these are the position of a 'Chief Information Officer (CIO)' as a member of the executive board in order to promote the integration and co-ordination of information provision and processing in all fields of the KIT at the top management level; and the 'Committee on Information Provision and Processing (IVA)', which was established by the senate and the executive board chaired by the CIO and gives recommendations to the president's office and the senate on all questions concerning the processing or provision of information at KIT, especially in matters of general importance. There are two sub-committees established by the IVA, one for information provision (A-IVS) and one for information processing (A-IFE). Two more will follow: one dealing with the 'application of new media in research, teaching, study, further education and administration (A-ILS)' and one for 'information processing in administration (A-IVD)' (Figure 2).

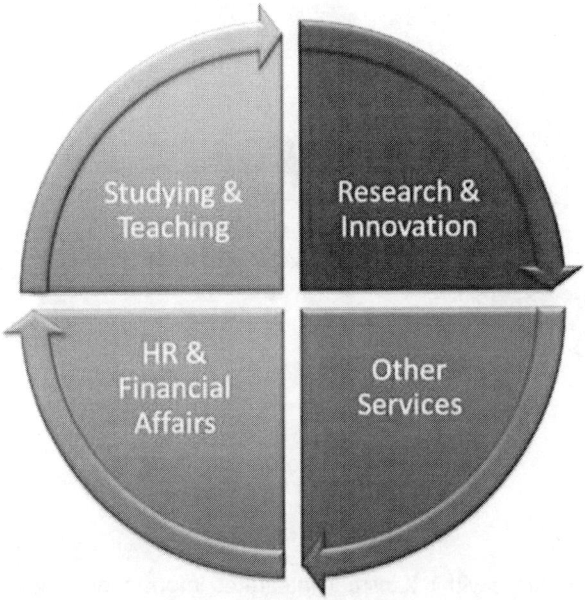

Figure 1. Project Portfolio within Karlsruhe Integrated Information Management (KIM).

Guidelines for these sub-committees form user governance: the sub-committees represent the interests of users in the respective fields, advise the operational officers and give appropriate recommendations for strategic development. Integration with the IVA is guaranteed by the fact that the respective chairmen of the subcommittees have to be members of the IVA. At the operational level, all relevant organisational units are represented in the 'Media and Information Service Centre Karlsruhe (MICK)'. MICK is organised co-operatively by the Steinbuch Centre for Computing, KIT Library and the administrative units dealing with information provision and processing. It pools all competencies and services in this domain. On the other hand, it does not break up or absorb the computing centre, library or administration as distinctive organisational units, as they have unique tasks and missions which do not overlap.

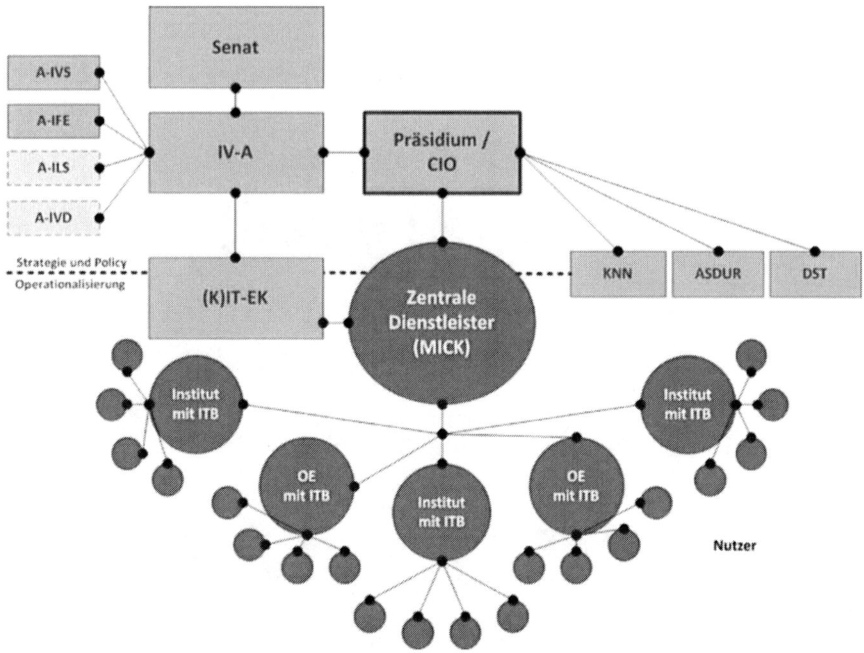

Figure 2. Governance within Karlsruhe Integrated Information Management (KIM).

CRIS AND REPOSITORIES

CRIS (Current Research Information Systems) have existed since the 1960s and have been designed to manage information related to research. This includes input and output data like publications, patents, products, prices as well as projects, equipment and facilities. At the European level, there exists a standardised data model called CERIF (Common European Research Information Format) which is maintained and developed by euroCRIS.[1]

Repositories were developed in the 1990s as part of the movement to secure open access to scholarly publications. Therefore they concentrated on providing free and unrestricted access to publications and not to the contextual data resulting from the research processes which led to these publications. As discussions and developments have matured, repositories have become much more aware of the importance of contextual information. The discussion now focuses on publication management, virtual research environments and research data.[2] Critics say this is due to the fact that, despite numerous initiatives and incentives,[3] repositories are not as well filled with full-text publications as promoters of the idea had hoped they would be.[4]

After initial discussions within the Knowledge Exchange (Razum et al., 2007), there is now a wider uptake of the question as to how and to what purpose CRIS and repositories should integrate or interact. Three main lines of discourse can be distinguished. Some think that a repository will develop into a CRIS (Carr, 2010), adding more contextual and standardised metadata about persons, projects, funders etc. The majority within both communities think that a parallel development will continue, and consequently there has to be some kind of interface between the systems depending on local requirements — be it one or bi-directional (Sheppard, 2010).[5] As a third perspective, some argue that a CRIS already has the ability to contain full-text publications and has to be enhanced accordingly to act as a repository. Their main argument is that a CRIS already has a rich and contextual data model which can express temporal roles and relationships between entities (Jeffery et al., 2010).

KIM-FIS — the Current Research Information System at KIT

Information about the research output and the context of research at KIT is currently held in numerous systems run by different organisational units using different formats and data models. It is not possible to combine, aggregate or integrate this rich information, whether it regards scientists, institutes or other characteristic features. This became all the more evident when the University and Research Centre Karlsruhe merged to form KIT. Both organisations had developed their own policies and daily practices in collecting, managing and evaluating research information.

These heterogeneous 'cultures' make it difficult if not impossible to gain a consistent overview of and insight into the research performance of KIT. However, such an overview is of crucial importance for KIT because it is competing with the top research institutions on an international level.

Therefore, in March 2010, the executive board of KIT initiated a project to build an integrated research information system as one of the sub-projects of KIM called KIM-FIS. This means that it will be part of the service-oriented concepts developed within iSOA. In a tendering process, KIT decided to choose Converis (by AVEDAS AG)[6] as a software platform for building its CRIS. Converis fulfilled most of the requirements which had been defined and both parties were interested in a strategic partnership promoting agile development on the one hand and long-term development of the product on the other. After the decision to acquire Converis in

July 2010, a demonstrator was built which focused on managing information about publications and projects. KIT library is responsible for the former, the research department for the latter. The project is co-ordinated by the CIO and there is a steering committee in which the library, research department and the department for planning and control are represented. The researchers and their activities are at the core of both sub-projects within KIM-FIS. A key challenge is to automate data collection and integration as much as possible. This includes migration of the existing publication databases, repositories and project databases within the former University and the former Research Centre, which are not integrated in researchers' workflows very well. In KIM-FIS, data is acquired and validated only once and then they can be used for manifold purposes, such as websites, CVs, publication lists and reference management software. In December 2010 the demonstrator was presented successfully to different stakeholder groups within KIT, namely researchers, the executive board and the Committee on Information Provision and Processing (IVA).

PUBLICATION MANAGEMENT

Publication management within KIM-FIS is also part of the open access strategy of KIT. As researchers strive for impact and visibility for their published results, they are willing to provide full-text files as long as they can be sure that copyright issues and other administrative burdens are taken care of. This is underpinned by a rigorous publication policy within the former research centre. It states that all publications have to be approved by the chair of the respective institute before they can be handed over to a publisher. The approval workflow ensures that coverage of the publication database of the former research centre is nearly 100%. The database has existed since 1956 and it holds about 75,000 records. At the former University, a publications database has existed on a voluntary basis since 1968 and it contains about 60,000 records. Currently a general publication policy for KIT is in the making which tries to combine the best of both worlds. The central idea is that the chair of an institute can decide if he/she wants to approve publications planned, as he/she is responsible for the scientific output of his/her unit. KIM-FIS will support the workflows which are currently being tested and rolled out to selected researchers and institutes. A first production release of the system was scheduled for August 2011. For the remainder of 2011, more institutes will gradually use KIM-FIS and give feedback, allowing for one or more iterations in development if necessary. The general roll-out of the system is scheduled for 2012.

From the very beginning, it became obvious that KIM-FIS was strongly linked to the identity management system already established in KIM-IDM.[7] It provides single-sign-on facilities for a wide variety of internal and external services at KIT. Roles and rights management must be clearly defined and the unique identification of researchers, funders and organizational units must be warranted. Internationally standardised persistent identifier systems for these and other entities are a natural requirement within projects like KIM-FIS, as they will greatly help the mobility of researchers' portfolios. However, there is not even a consensus about which persistent identifier systems should be deployed on a global scale for publications and which services should be supplied by them (Hakala, 2010). ORCID[8] and other activities in this domain are carefully monitored; for the time being a pragmatic 'bag of identifiers' approach seems most promising for KIT. This means that maybe more than one (persistent) identifier will be attached to a person or a publication.

In addition to the publication databases, two open access repositories have been developed locally, one at the former University and one at the former Research Centre Karlsruhe. KIT is a strong actor in open access. It has passed an institutional policy endorsing the green and gold routes. It has signed the Berlin Declaration on Open Access to Knowledge in the Sciences and Humanities and has established a fund to support open access publication charges.[9] It also runs KIT Scientific Publishing,[10] an institutional open access publishing house which actively engages in the Association of European University Presses (AEUP).

It soon became obvious that none of the locally developed repositories would be the long-term solution to supporting the green road to open access at KIT. Two options remained: to migrate to a new repository platform like dspace or fedora, or to expand the possibilities to handle full-text publications within Converis. As has been said before, one of the main arguments in favour of full integration is that Converis already supports a rich and contextual data model which can express temporal roles and relationships between entities. Therefore, it seems natural to follow an integrated approach which avoids the need to develop and maintain interfaces between the CRIS and a repository.

As KIM-FIS is gradually being deployed at KIT during the second half of 2011, a separate project (KIM-FIS Repository) will develop additional functionality to turn Converis into an open access repository. The benchmark will be the DINI Certificate for Document and Publication Services (Dobratz and Scholze, 2006).[11] DINI, the German Initiative for Networked Information, is a coalition formed by German Higher Education infrastructure and service institutions, such as libraries, computing centres and media centres, as well as by scientific learned societies. The DINI Certificate aims at networking document and publication repositories by promoting the use of standards, interoperability and co-operation between Higher Education institutions. It formulates requirements and recommendations covering technical, organisational and policy aspects like visibility, author support, legal issues, indexing and interfaces, access statistics and long-term availability. Looking at the technical parts of the DINI Certificate, the main focus in development will be on external interfaces, like OAI-PMH, SWORD, OpenAire and others, adding to the visibility of publications which is not a traditional feature of a CRIS. From a researcher's perspective this makes perfect sense. A research information system is the back end in which his/her whole research portfolio is available for re-use and external communication. It should provide publication lists for his/her website, for project proposals as well as for his/her CV or for scholarly communication with peers. From this perspective, separating a CRIS from a repository seems is much like erecting an artificial barrier, although the historical development is clear. A CRIS will always have an internal side which is used for evaluation and reporting. This does not prevent its rich and networked data from being displayed externally to achieve maximum visibility. When it comes to publications, it enables storage of all full-text articles and then display of links to the publisher website or internal or external repositories depending on the publisher's open access policy and the user's rights and roles (Figure 3).

Figure 3. KIM-FIS demonstrator with links to the published and the repository version of a publication.

PERSPECTIVES

The motto for developing KIM-FIS is 'think big in small steps'. Therefore a CERIF-based rich data model has been implemented, although, as first steps, only the publication and project information modules will go into production service. Patents, prices and other assets of research output will follow later in 2012 and 2013. By this time, publication management including repository functionality will be operational for some time and will play its part in a large-scale evaluation exercise within the Helmholtz Association of Research Centres, the so-called Programme Oriented Funding (POF).[12] Indicators based on publications will be calculated for comparing the Research Centres (KIT being one of them). KIM-FIS will then be the backbone for this exercise at KIT.

ACKNOWLEDGEMENTS

The authors would like to thank Karlsruhe Institute of Technology (KIT) and the Ministry for Science, Research and the Arts of Baden-Württemberg (MWK) for supporting this work.

REFERENCES

1. Björk, B-C., P. Welling, M. Laakso, P. Majlender, T. Hedlund, et al. (2010): 'Open Access to the Scientific Journal Literature: Situation 2009'. PLoS One, 5(6), e11273.

2. Carr, L. (2010): 'EPrints: A Hybrid CRIS/Repository'. In: Workshop on CRIS, CERIF and Institutional Repositories, 10–11th May 2010, Rome, Italy. http://eprints.ecs.soton.ac.uk/21048/.

3. Dobratz, S. and F. Scholze (2006): 'DINI Institutional Repository Certification and Beyond'. In: Library Hi Tech. pp. 583–594.

4. Freudenstein, P., L. Liu, F. Majer, A. Maurer, C. Momm, D. Ried and W. Juling (2006): 'Architektur für ein universitätsweit integriertes Informations- und Dienstmanagement′. In: Proceedings of INFORMATIK 2006 — Informatik für Menschen, 36. Jahrestagung der Gesellschaft für Informatik. Dresden. pp. 50–54.

5. Gaedke, M., J. Meinecke and M. Nussbaumer (2005): 'Aspects of Service-Oriented Component Procurement in Web-Based Information Systems'. International Journal of Web Information Systems (IJWIS), 1(1), 15–24.

6. Hakala, J. (2010): 'Persistent identifiers — an overview'. KIM Technology Watch Report 2010. http://metadaten-twr.org/2010/10/13/persistent-identifiers-an-overview/.

7. Hartenstein, H., W. Juling and A. Maurer (2007): 'Karlsruhe Integrated Information Management — KIM'. In: Changing Infrastructures for Academic Services. Information Management in German Universities. Göttingen: DINI — Deutsche Initiative für Netzwerkinformation, http://www.dini.de/fileadmin/docs/DINI_Informationsinfrastrukturen.pdf.

8. Jeffery, K.G., A. Asserson and D. Luzi (2010): State of the Art and Roadmap for Current Research Information Systems and Repositories. euroCRIS White Paper. http://www.irpps.cnr.it/it/system/files/PositionPaperRoadMap20100610.doc.

9. Razum, M., E. Simons and W. Horstmann (2007): Institutional Repository Workshop Strand Report — Exchanging Research Information. http://www.driver-repository.be/media/docs/KEIRstrandreportExchangingResearchInfoFINALFeb07.pdf.

10. Sheppard, N. (2010): 'Learning How to Play Nicely: Repositories and CRIS'. In:Ariadne Issue 64 http://www.ariadne.ac.uk/issue64/wrn-repos-2010-05-rpt/.

NOTES

[1] http://www.eurocris.org

[2]See the talks at the DINI-Helmholtz-Workshop: Repositorien — Praxis und Vision, Berlin 2010, http://www.dini.de/veranstaltungen/workshops/dinihelmholtz-workshop-repositorien-praxis-und-vision/programm/

[3] http://ec.europa.eu/research/science-society/index.cfm?fuseaction=public.topic&id=1294&lang=1

[4](Björk et al., 2010) show that repositories in the STM field cover about 12% of the published Journal articles. Cf. Paul Jump: Research intelligence — Empty buckets or pure gold? In: Times Higher Education 17 March 2011.

[5]Workshop on CRIS, CERIF and Institutional Repositories. 10–11th May 2010, Rome, Italy. http://www.irpps.cnr.it/it/eventi/workshop-on-cris-cerif-and-institutional-repositories. 2nd WORKSHOP ON CRIS, CERIF AND INSTITUTIONAL REPOSITORIES 'Integrating Research Information: CRIS+OAR' 23–24th May 2011, Rome, Italy. http://www.irpps.cnr.it/it/eventi/2nd-workshop-on-cris-cerif-and-institutional-repositories-integrating-research-information-crisoar.

[6] http://www.avedas.com/en/converis.html.
[7] http://kim.cio.kit.edu/138.php.
[8] http://www.orcid.org.
[9] http://www.bibliothek.kit.edu/cms/english/publishing.php.
[10] http://www.ksp.kit.edu.
[11] http://www.dini.de/dini-zertifikat/english/.
[12] http://www.helmholtz.de/en/about_us/programme_oriented_funding/.

Chapter 7

Aligning Web Content and Organisational Strategy: Towards Increasing Funding

Khutso Maahlo[1], Molefe Ratsoana[1], Martie A. Mearns[2]

[1]*Centre for Information and Knowledge Management, University of Johannesburg, South Africa*

[2]*Department of Information and Knowledge Management, University of Johannesburg, South Africa*

ABSTRACT

Background: The effectiveness with which the Aurum Institute manages to communicate its organisational strategic objectives through its website was investigated. Despite its ground breaking research and programs which have made a positive impact on the community, the major problem that it faces is funding its core.

Objectives: An investigation of the website's content was carried out to determine the extent to which the organisational strategy is reflected to potential funders visiting the site. Requirements for aligning the content with the organisational strategy were identified.

Method: Content analysis was used where secondary data, such as website content and organisational strategic objectives, was analysed. An interview was also conducted with Aurum's knowledge manager to gain an in-depth understanding of the organisation's strategic objectives and to validate the initial findings.

Results: The results show that the website content was not effectively representative of Aurum's strategic objectives, and that the website structure did not effectively share the information that supports the decision-making process of potential investors. Recommendations were presented to the organisation in the form of a prototype website which reflects the desired website content that is representative of Aurum's strategic objectives.

Conclusion: Further research is required in determining the effect that a strategically driven website could possibly exert on the funding potential of the organisation. The theoretical approach used in this study can be used as a scoping exercise in organisations of a similar business nature.

INTRODUCTION

Website content that is effectively integrated with the strategic objectives of an organisation uncovers opportunities of increasing its audience base, specifically in the case where the website is geared towards drawing the attention of potential investors or funding opportunities. According to Akakandelwa (2011:421), websites act as connection and communication points for information users with electronic information. As a result, most corporations, organisations or institutions are making efforts to launch themselves into the virtual world of the World Wide Web. Rocha (2012:374) explains online perspective of organisations today, that they: '(I)nvest time and money to develop and maintain their website's quality. These websites should establish an effective information and communication channel between organisations and their clients'.

Panopoulou, Tambouris and Tarabanis (2008) note the following:

Literature suggests that the sophistication of websites is related to a number of characteristics and features regarding interface and aesthetics, navigation, consistency and patterns, content relevancy and accuracy, accessibility, security and currency of information. (p. 518)

However, these characteristics and features need to 'be reviewed regularly to determine whether they are fulfilling the reasons for which they were developed' (Chiou, Lin & Perng 2011:1463).

It has been identified that, despite the recognised importance of having a website, most organisations do not plan properly for it. Nacar and Burnaz (2011:276) and Liao, Proctor and Salvendy (2008:44) highlight that 'content preparation could be handled from two main perspectives: the information content to be provided and the structure of how this information is to be presented'. The information content and structure need to be addressed in the website strategy of the organisation. Organisations also make use of website strategies to communicate and collaborate with stakeholders and inform them of the products and services that the organisation must provide (Chiou, Lin & Perng 2010:286). The website strategy development process ensures that an organisation addresses the strategic objectives that the website should achieve, one of which should be to effectively communicate its vision, mission and objectives, together with evidence that supports these aims, to its users.

It has become increasingly important that an organisation operates from a business strategy that ensures the efficient allocation of resources to relevant tasks. This is even more important for non-profit organisations (NPOs) because their funding comes from external sources. The same principle applies to their website strategies. NPOs are required to effectively communicate their mission and vision to the users of their website by certifying that their strategic objectives are well communicated. It is the purpose of this article to report on research conducted at the Aurum Institute. The Aurum Institute is an NPO based in South Africa whose aim is to improve on the health of the community through innovation in Tuberculosis and HIV and AIDS research (Aurum Institute 2012:3). Prof Gavin Churchyard, CEO of the Aurum Institute stated in the annual report of 2010 that 'one of the major problems that face us is funding our core' (Aurum Institute 2010:45). Aurum owns a website on which interested parties and potential funders can access information about the institute and its research initiatives. In addressing the above-mentioned funding issue, Aurum's website content was assessed to establish if it is effectively aligned to the strategic objectives of the organisation.

An organisation can make use of a website as a medium which will be accessed to transfer information and knowledge, and facilitate positive outcomes of the decision-making process made by the investors (Bou-Llusar & Segarra-Cipres 2006:106). The website designers need to analyse the content using relevant analysis tools to ensure that it is of sound quality, relevant, timely and that it facilitates decision-making. The organisation needs to identify the content it wishes to present on the website which will appeal to the targeted audience. Bill Gates (1996) also emphasised the power of persuasion that content possesses. He states that the quality and presentation of the content has the power to lure or push the audience away. In this context it is argued that *a conscious inclusion of organisational strategy into website content is required to enhance Aurum Institute's credibility, strengthen its organisational relations and increase its competitive advantage for better funding opportunities.*

This argument is based on the findings generated through the investigation that addressed the following main research question:

How effective is the Aurum Institute's website content in portraying their strategic objectives towards potentially attracting external funding or sponsorships?

In order to have addressed this research question the study was guided by the following sub-questions:

- What is the content quality and condition of the current website of the Aurum Institution?
- What are the organisational goals, objectives, values and strategy of the Aurum Institution and how are they currently represented by the website?
- What needs to be carried out to ensure that the Aurum Institute's content is aligned with its objectives?
- Using the findings from the above sub-questions, what website content will be suitable for the implementation of a website that will represent organisational goals and attract investments?

A qualitative approach was followed to collecting and analysing data, collecting and analysing secondary data as well as to conducting an in-depth interview. The secondary data comprised of Aurum's knowledge audit that was completed a year prior to the commencement of this study, and the inclusion of the current website content and the organisation's strategic objectives. A structured interview took place with the knowledge manager of Aurum Institute to fill the data gaps identified. All results were analysed and recommendations derived and integrated in the new prototype website.

LITERATURE REVIEW

Dias and Reinhard (2004:5) maintain that organisations can add value to the business processes and activities by understanding the importance of Internet technologies and their platforms. Bailey (2010) mentions Gates (1996) who claimed that organisations could facilitate the management of information and the creation of quality website content, which will result in added value to the processes, procedures, communication and collaboration channels of those organisations. Chiou *et al.* (2010:286) stated that implementing a website strategy within the organisation can increase communication and collaboration with its stakeholders and customers. Moustakis, Tsironis and Litos (2006:60) identified the important factors that characterise websites of good quality; these include the following:

- content
- navigation
- structure and design
- look and feel
- multimedia used.

Huizingh (2000:124) emphasised the importance of the structure and quality of content as a result of the vital role they play in persuading stakeholders and customers.

The nature of Aurum's business necessitates that the organisation attracts potential investors to ensure that projects are well resourced and are effectively executed. De Troyer (1998:2) identified that users experience problems with organisations' websites; these include, amongst others, unclear mission statements of the organisation, no clearly defined target market and poor management of content. An organisation can increase the satisfaction levels of its stakeholders and increase revenue through effective information and knowledge sharing with them (Mpindiwa 2010:4). According to Hernandez, Jimenez and Martin (2009:362) a large number of studies in the 1990s placed much emphasis on website design (the content, use and objective of the website) but failed to identify the factors that improve the organisation's website strategy. It is imperative that the organisation understands the website strategy as well as its purpose in the organisation's communication plan before it is integrated.

WEBSITE STRATEGY

According to Hernandez *et al.* (2009:364) organisations engaged in electronic commerce aided by an effective website strategy have the potential of being exposed to a larger market than organisations that do not effectively incorporate website strategy. Schmidt, Cantallops and Dos Santos (2008:504) stated that the implementation of the website strategy in the organisation enhances communication with stakeholders, and improves marketing mix and business practices. The website strategy is utilised by individuals or organisations to inform their audiences about the services and products offered by the organisation (Chiou *et al.* 2010:282). As part of their website strategy it is important that organisations identify the characteristics that enhance the interaction and relationship of the organisation with its stakeholders and customers (Baloglu & Pekcan 2006:174). Hernandez *et al.* (2009:364) identified these characteristics as the factors that are inherent in the Web Index Assessment (WIA) which facilitate effective and efficient access to the organisation's information through the website. Below is a brief summary of the factors that improve the quality of a website:

- **Accessibility**: this term refers to the ability of potential and current users to identify and access the website and its content. The volume of traffic that reaches the website influences the accessibility of the content on the website. There are two indicators that measure accessibility to the website; these are the position of the website on the search engine results and the popularity of the website. Search Engine Optimisation (SEO) and Search Engine Marketing (SEM) are tools that are used to measure the accessibility of the website. The SEO tool works to improve the traffic of the website using the search engines by submitting targeted keywords. The SEM tool works to obtain free search listings from the search engines by purchasing the search listings. It aims at increasing the traffic by improving the website's visibility on the search engines.

- **Speed**: this factor plays a vital role in the satisfaction levels of users and facilitates effective transactional processes on the virtual space. Website designers need to ensure that the loading time of the website is fast, and failure to do so will lead to a potential loss of users searching for information to other websites. Websites with a slow loading time increases the possibility of pushing potential customers away. Hernandez *et al.* (2009:364) stated that if a website takes more than 10 seconds to load, the website has a high chance of losing the user.

- **Navigability:** website designers need to put in place applications that will ensure ease of use and ability to find the content on the website. Poorly designed websites may result in inaccessibility to valuable information which could have the potential of dissuading stakeholders or customers to invest in or purchase into the organisation. Hernandez *et al.*(2009:364) further identifies the factors that website designers should consider in the design of the website which will enable users to successfully navigate through the website. Firstly, the content should be classified and organised according to a relevant topic and headings; secondly, the information should be labeled; thirdly, the navigation system should be designed in such a way that it will guide the user through the content and fourthly, the users must be guided to the information they need in order to make decisions.
- **Content quality:** the basic goal of implementing a website strategy is to provide information that is relevant to the organisation. There are two functions, or roles, that content plays on the website. (1) It informs the audience about the organisation's background, the products and services it offers, and the vision of the organisation. The information should be regularly updated and relevant. It should also be precise to ensure that a user has a full understanding of the organisation without being subjected to information overload. (2) It facilitates communication and interaction between the organisation and the customers. Website designers need to create a communication medium that will be used to share content with the customers and also interact with them (Hernandez *et al.* 2009:364).

By identifying the above characteristics an organisation can determine the website features that are important to incorporate into its website strategy.

Grigoroudis *et al.* (2008:1346) explain the following about quality and communication: Organisations invest time and money to develop and maintain their website's quality. These websites should establish an effective information and communication channel between organisations and their clients.

Hussainey and Mouselli (2010:6) maintain that stakeholders need to be well informed in order to make good decisions regarding purchases or investments in the organisation. Stakeholders are persuaded by the aesthetics and the manner in which the content is displayed on the website when they decide about investing in the organisation (Hussaney & Mouseli 2010:6). Huizingh (2000:124) claims that the engagement of stakeholders with the organisation is dependent on the information shared by the organisation, together with its ability to communicate the potential growth and return on investment of the organisation. Organisations that are non-profit in nature depend on financial assistance to successfully implement their projects. It is therefore of the utmost importance that the information they share with their stakeholders, through their website, communicates their plans for growth and the potential return on investment.

Porter (2001:64) recognises the importance of considering corporate strategies when designing the website strategy. He identifies principles that organisations could use as guidelines when positioning the website strategy in the organisation to ensure that it complements the corporate strategies. Gakhar (2012:74) states that an effective alignment of the website strategy to the organisational strategies increases investment opportunities for the organisation and potential customers. The alignment plays a vital role in the decision-making process of the potential investor and determines the outcome of the decision-making process.

INVESTMENT CRITERIA USED BY INVESTORS AND FUNDERS

According to Kollman and Kuckertz (2010:5) there are investors who raise money and then look for an organisation that has the potential to yield a high return on their investment. The investors engage in investment processes where the deal origination is determined. The deal is screened, evaluated and the structure of the investment is analysed to identify whether or not the structure will support the execution of the project at hand (Kollman & Kuckertz 2010:9). Once the investment decision has been made, the investor will monitor the performance of the organisation to ensure that the goals and objectives that were set are being executed effectively. The organisation is constantly monitored to determine whether it is producing the desired results or not. There is an additional set of investment criteria which will be discussed in the ensuing section.

When investing in organisations, an investment criterion is used to evaluate the potential of the organisations that seek investment opportunities. Kollman and Kuchertz (2010:27) have identified the factors that investors consider before investing in an organisation or providing funds that will facilitate for completion of the organisation's projects. Below is a discussion of these factors:

- *Personalities of the entrepreneurs:* the skills, capabilities and competencies of the leaders of the organisation are assessed based on the achievements of the organisation under their leadership. Leaders play influential roles in the behaviours and attitudes of employees by setting the culture that filters throughout the organisation. The business practices that are employed by the leaders are assessed to see whether or not they are ethical and whether they are efficient in achieving the goals and objectives or not. The reputation of leaders is also assessed to reduce risks of losing returns as a result of the leader reputation-factor.

- *Experiences of entrepreneurs:* investors assess the organisation's growth and the development that has taken place over a certain period of time by looking at its track record. There are activities and business practices that require organisations to be accredited or qualified for practice. The investor evaluates the business qualifications and the technical qualifications of the organisation and its staff to ensure they have the relevant experience.

- *Products or services:* the needs of customers continuously evolve and require organisations to be flexible, innovative and responsive to those needs. Investors assess products and services and identify how the organisation adds value to the final product. The organisation claims its competitive advantage from its unique selling point, core competencies and capabilities which are assessed to identify whether or not they will be able to sustain the innovation rate.

- *Market characteristics:* the potential risks and threats that reside in the market of the target investment are identified and assigned weights to calculate the potential impact of a threat. The potential growth of the market is also assessed by measuring how the market accepts new products and services. The market players and external factors that influence the operation of the organisation are also identified and assessed.

- *Financial characteristics:* the investor assesses the potential return on investment when considering an investment in the specific organisation. The financial characteristics of the organisation determine whether or not the investor will invest in the organisation or will seek other investment opportunities (Kollman & Kuchertz 2010:27).

The above-mentioned investment criteria are factors that were considered when the strategic objective measurement indicators were developed to assess Aurum's website.

There are different types of investors, which include, firstly, international investors who are located in a different country from where they invest. Secondly, the other type of investor is the domestic investor who resides within the same country in which the investment lies (Kollman & Kuchertz 2010:15). International and local investors who seek investment opportunities often make use of Internet technologies to access different websites of organisations that seek investment opportunities. The content quality, the accessibility, navigability and the downloading speed of the website will determine whether the investor is more likely or not to invest in the organisation. Organisations make use of websites as a medium that can transfer information and knowledge which can facilitate positive outcomes for the decision-making process followed by the investors (Bou-Llusar & Segarra-Cipres 2006:102).

METHODOLOGY

The Aurum Institute was used as a case study to examine the factors that determine the quality of a website's content and how information and knowledge is shared by the organisation with its customers and stakeholders. Secondary data was analysed which included the outcome of Aurum's knowledge audit (which was performed one year prior to this study by Aurum's knowledge manager), Aurum's website content as well as their organisational strategy document. Liebowitz *et al.*(2000:3) defines a knowledge audit as a tool appropriate to evaluate how the organisation would need to be changed to better facilitate for the flow and sharing of information and knowledge. The website content and the strategy document were compared to identify any discrepancies that existed between the two data components.

Data was collected using an observation schedule to assess the characteristics of Aurum's website and to evaluate the content. As a result of the dynamic nature of the website, where regular changes are made, the content was captured, for a period of one week, for website observation. This approach is in line with the principles of 'one–way–mirror' research to allow the analysis of website content and secondary data without any alterations of the content (Healy & Perry 2000:119). This approach allowed the authors to provide a description of what was experienced and measure that which was observed (Krauss 2005:760).

A qualitative approach to collecting and analysing data was employed to produce the findings for the research problems identified. The qualitative method of collecting data provided the authors with descriptive data, which facilitated processing and the unearthing of findings that are in-depth and descriptive. This type of research is empirical, according to Wacker (2008:10), because the study's expected results or findings are used as inputs into the designing processes of the prototype website, which will represent the desired website strategy. Secondary data were collected for analysis, which included Aurum's organisational strategy document as well as their website content. The website was analysed and compared to the organisational strategies to identify the gaps. After the analysis of the secondary data, an interview with the knowledge manager was conducted. This was to allow for the validation, triangulation and clarification of the data findings and insight regarding the assessment of the website content against the organisational strategy. The interview also clarified matters for the development of a prototype website. The nature of the data prescribed the descriptive analysis techniques for the analysis of secondary data.

Coding was used to analyse the secondary data qualitatively. This analysis was applied to identify the associations between the different aspects of the data or codes using a relationship analysis technique, as identified by Saldaña (2009:3). The collection and analysis of data were performed with the aim of creating a website prototype that will successfully compliment the organisational strategic objectives of Aurum, and potentially increase investment opportunities. The organisational strategic objectives were translated into measurement indicators, and used as benchmarks to measure how each webpage on the website represented each of the identified objectives.

Ethical practices were considered in the collection and analysis of data. The Aurum Institute's strategic objectives that were used in the secondary data analysis, are confidential and have strategic value to the organisation. Disclosing this information may jeopardise the success of the organisation and its projects and therefore the data analysis used undisclosed codes. In the interview session the information of the participant remained confidential, and permission was requested from the organisation to conduct the research. The highest technical skills and methodologies were employed in handling the data to avoid fabrication of the data.

Limitations that were encountered included the fact that Aurum's website content was subject to change at any given moment. The results of the content analysis only apply to a specific period of time as the website content has since been updated, edited or removed. However, the lessons learnt from the findings remain applicable from a holistic perspective regardless of editorial and maintenance updates.

Findings: Website content representative of organisational goals

Descriptive analytical tools were used to identify the patterns and the relationships between the organisational website and the corporate strategic objectives from the identified sources of secondary data.

The results, from the analysis, were used as inputs in the designing and development of the website prototype. The first prototype presentation was based on a gap analysis that resided between the website and the organisational strategic objectives.

The strategic objectives were translated into measuring indicators against the website to identify how each webpage represents the strategic objectives. Table 1 illustrates how each objective was represented by each of the webpages.

After the development of the first prototype, an interview was conducted with Aurum's knowledge manager and aligned to the knowledge audit that was conducted prior to this study. This interview also triangulated the requirements necessary to successfully represent the organisational strategic objectives. The feedback received from the interview was incorporated into the re-designing stage of the prototype which was then used as recommendations to the organisation.

Table 1 illustrates the analysis of Aurum's website content against the organisational strategic objectives that were used as the measuring indicators. The webpages were individually assigned weights using a three–

TABLE 1: Representation of organisational strategic objectives by each webpage

Webpage visited	Objective						Total objectives per page
	1	2	3	4	5	6	
Home page	3	2	1	3	3	3	15
Management team	1	1	1	3	1	3	10
Aurum in the news	3	1	1	3	1	1	10
Aurum events	3	3	1	3	3	3	16
Aurum staff awards	1	1	1	1	1	1	6
Career opportunities	1	1	1	1	1	3	8
Aurum training	3	2	1	3	1	3	13
Scholarships	1	1	1	3	1	3	10
Publications	1	3	1	1	1	3	10
Funders	1	1	1	3	1	1	8
Collaborators	1	1	1	3	3	1	10
Contact us	1	1	1	2	1	1	7
Total representativeness of website	-	-	-	-	-	-	**123/213**

Keys: 1 = Hidden; 2 = Obscure; 3 = Visible

point scale which determined whether or not the strategic objectives were (1) not effectively represented or hidden, (2) obscure or ambiguous, or (3) visible and clear. The highest total that a webpage could score was 18. The maximum possible weight of 213 was calculated by multiplying the six objectives with the highest possible weight (3 – visible) that the webpage could score per objective. Each webpage was given a total weight out of a possible maximum of 18 (webpage weight is the sum of all the points attained from each objective per webpage). The results derived from the analysis of the findings presented in Table 1 are subsequently discussed:

- **Home page:** This page was allocated a score of 15 out of 18, meaning it is 83% representative of Aurum's strategic objectives. This webpage achieved this weighting score by including information regarding disease management with specific focus on prevention, treatment and care of the communities. The programs that the institute engaged in were also covered and included on the home page. The home page is the information users' first point of contact with the organisation's content, therefore a score of 83% indicates that it is more representative of Aurum's strategic objectives. This high representativeness enables the information user to obtain a clear understanding of the organisational strategic objectives and gives funders an idea of projects they could potentially sponsor.

- **Management team:** This page was allocated a score of 10 out of 18, meaning it is 56% representative of Aurum's strategic objectives. This was achieved by including information about the staff responsible for leading Aurum. Information on their professional qualifications was also made available. This information enables the potential funder to gain insight in the credibility of the management team that leads the organisation.

- **Aurum in the news:** This page was allocated a score of 10 out of 18, meaning it is 56% representative of Aurum's strategic objectives. This was achieved by the page's inclusion of the various news articles and information on community work carried out around the country. This information is vital as it could assist potential investors to assess the publicity and reputation of Aurum.

- **Aurum events:** This page was allocated a score of 16 out of 18, which indicates that it is 89% representative of Aurum's strategic objectives. This representation was achieved by the page's inclusion of information about the events that Aurum has hosted and projects it undertook. This information is relevant for potential investors as they are able to see the outcomes of investments made by other funders. This thus gives them an idea of how their funding would be spent should they decide to invest in Aurum.

- **Aurum staff awards:** This page was allocated a score of 6 out of 18; this shows that it is 33% representative of Aurum's strategic objectives. This was achieved through the page's inclusion of information on the awards that were given to Aurum's staff who performed exceptionally well. Although this information adds value, and could be useful for potential investors by enabling them to identify the skills and abilities of the Aurum staff, this page should be on the institute's intranet and not the website. This is because it is more focused on internal staff affairs and internal recognition of performance.

- **Career opportunities:** This page was allocated a score of 8 out of 18, meaning it is 44% representative of Aurum's strategic objectives. This was evident in the page's inclusion of information on the vacancies that are currently available at the Aurum Institute, together with a description of the vacant positions. The 44% is understandably low, resulting from the nature of the page.

- **Aurum training:** This page was allocated a score of 13 out of 18, meaning it is 72% representative of Aurum's strategic objectives. This was accomplished by the page's inclusion of information on the clinical training, counselling training courses, seminars, bookings for training, downloads and past programs. This information is very important for potential investors as it provides insight into the quality of the curriculum offered by the Aurum Institute.

- **Scholarships:** This page was allocated a score of 10 out of 18, meaning it is 56% representative of Aurum's strategic objectives. This was achieved by including information on the scholarships provided by the organisation, which demonstrates a corporate social responsibility on Aurum's part, through provision of education or skills development programs. This attempt by Aurum to empower the community (corporate social responsibility) is likely to attract potential investors.

- **Publications:** This page was allocated a score of 10 out of 18, meaning it is 56% representative of Aurum's strategic objectives. It included information on topics of interest including the prevention, treatment and control of the HIV and TB pandemic. This page's representativeness percentage should be higher given that Aurum is a health research institute. Together with the research available, information should be provided on how this research will help Aurum in eradicating the scourge of TB and HIV and AIDS.

- **Funders:** This page was allocated a score of 8 out of 18, meaning it is 44% representative of Aurum's strategic objectives. It contained information on the funders who have worked with the organisation. This representativeness percentage should be higher as this information should give potential investors the confidence and security to dedicate resources to the Aurum Institute.

- **Collaborators:** This page was allocated a score of 10 out of 18, meaning it is 56% representative of Aurum's strategic objectives. This was achieved through the inclusion of information on those organisations, both international and domestic, that the Aurum institute works with to successfully improve the welfare of communities. This page's representativeness should also be higher, given that it is one of the pages that are most likely to be visited by potential investors.

- **Contact us:** This page was allocated a score of 7 out of 18, meaning it is 39% representative of Aurum's strategic objectives. It included Aurum's contact details at head office as well as the addresses of the different branches. The nature of this page allows it to score in this percentage as it is generally not a content heavy page.

Based on the results presented above, it is now known that Aurum's website content, as a whole, is 57.7% representative of the institute's strategic objectives. The percentage was calculated by dividing the sum total of all the webpages (123) with the maximum total that the webpage could score (213). From the measurement indicators used on the individual webpages, it was illustrated how the website achieved this percentage. A result of 57.7% indicates that Aurum's website content does not successfully communicate the organisational strategic objectives. The analysis of the results above also offers a clear indication of the webpages on which the organisation's objectives should have or have been displayed.

CONCLUSION

The significance of this study was to establish the extent to which Aurum's website content is aligned to the organisation's strategic objectives. Based on the research findings it is evident that Aurum's website content is not effectively representative of these strategic objectives. The reason for this could be attributed to information overload, insufficiency in relevant information and the presentation of content. These issues could be negatively affecting the funding, causing the difficulties that Aurum currently faces. It has been noted, however, that some of the webpages are not meant to contain too much information and, thus, contributed to the low percentage that Aurum scored in its overall representativeness of the strategic objectives.

The theoretical approach that was taken can be utilised to assess the websites of other organisations that aim to achieve similar results. The results from this study are appropriate for application to other NPOs, as well as for–profit organisations, to analyse their website content. During the literature review phase of this study a shortage of academic research, on the importance of website content alignment to the strategic objectives of an organisation, was realised. Further research must be conducted on this topic, particularly for non-profit organisations, as this has an impact on the potential funding that they could attract. Additional research will need to be conducted on the measurement of the funding that the Aurum Institute would attract, once their website content has been aligned to the strategic objectives of the organisation. In other words, two types of website should be measured and compared for how effective each is at attracting funding or sponsorships, one which has website content that is aligned to its organisational strategic objectives, in comparison to one that is not thus aligned. This would require a longitudinal research approach.

The processes of website planning and defining the purpose of the website content should be largely influenced by the organisational strategic goals of the Aurum Institute, to ensure that relevant content is placed on the website. Aligning Aurum's website content with the strategic objectives of Aurum will enable the organisation to effectively communicate information that is relevant to the target audience (potential investors), and will successfully facilitate quick and positive decision-making by the potential investor. The factors that constitute a successful website strategy include the loading speed, navigability, ease of access, and available content. These factors are most likely to influence the decision-making processes of stakeholders and potential investors. Further research opportunities exist to identify a model that will evaluate the information needs of investors.

This study provides Aurum with insight, on the specific aspects of their website that are currently lacking, for improved communication of its organisational objectives. The analysis shows the webpages where the organisation's objectives are visible, and where they are not, but should be visible. Furthermore, the website can now be developed into an effective platform that effectively reflects the essence of the organisation, in order to specifically gain the attention of potential funders who would need to know whether or not their funding intent is directed at the correct funding recipient.

It can be concluded that when a website's content is aligned to the strategic objectives of the organisation, and does so on a website structure that is easy to navigate and presents the information in an easily consumable manner, the website could be effective in drawing the attention of the user, which could ultimately translate into increased and improved funding.

ACKNOWLEDGEMENTS

Competing interests

The authors declare that they have no financial or personal relationship(s) that may have inappropriately influenced them in writing this article.

AUTHORS' CONTRIBUTIONS

K.M. (University of Johannesburg) and M.R. (University of Johannesburg) conducted the research as postgraduate students under the supervision of M.M. (University of Johannesburg). M.M. conceptualised the project and K.M. and M.R. collected and analysed the relevant data with supervision support from M.M.

REFERENCES

1. Akakandelwa, A., 2011, 'An exploratory survey of the SADC e-government website', *Library Review* 60(5), 421–431.

2. Aurum Institute, 2010, 'Annual Report 2010', viewed 18 August 2012, fromhttp://www.auruminstitute.org/phocadownload/annual_report_2010.pdf

3. Aurum Institute, 2012, 'Company profile', viewed 04 August 2012, from http://www.auruminstitute.org/publications.php

4. Bailey, C., 2010, 'Content is king by Bill Gates', viewed 12 September 2012, from http://www.craigbailey.net/content-is-king-by-bill-gates.html

5. Baloglu, S. & Pekcan, Y.A., 2006, 'The website design and internet site marketing practices of upscale and luxury hotels in Turkey', *Tourism Management* 27, 171–176.

6. Bou-Llusar, J.C. & Segarra-Cipres, M., 2006, 'Strategic knowledge transfer and its implications for competitive advantage: an integrative conceptual framework', *Journal of Knowledge Management* 10(4), 100–112.

7. Chiou, C.W., Lin, C. & Perng, C., 2010, 'A strategic framework for website evaluation based on the review of the literature from 1995–2006', *Information and Management* 47, 282–290.

8. Chiou, C.W., Lin, C. & Perng, C., 2011, 'A strategic website evaluation of online travel agencies', *Tourism Management*32(6), 1463–1473.

9. De Troyer, O., 1998, *Designing well-structured websites: Lessons to be learned from database schema methodology*, Tilburg University, Tilburg.

10. Dias, I. & Reinhard, N., 2004, 'Methodology for web presence strategy formulation: A case study', Proceedings *17th Bled e-Commerce Conference*, Slovenia, 21–23 June, Brazil.

11. Gakhar, D.V., 2012, 'Perception of stakeholders on web-based corporate reporting practices', *Journal of Advances in Management Research* 9(1), 64–76.

12. Gates, B., 1996, 'Content is king', Wordpress, San Francisco.

13. Grigoroudis, E., Litos, C., Moustakis, V., Politis, Y. & Tsironis, L., 2008, 'The assessment of user-perceived web quality: application of a satisfaction benchmarking approach', *European Journal of Operational Research* 187(3), 1346–1357.

14. Healy, M. & Perry, C., 2000, 'Comprehensive criteria to judge validity and reliability of qualitative research within the realism paradigm', *Qualitative Market Research – An International Journal* 3(3), 118–126.

15. Hernandez, B., Jimenez, J. & Martin, M.J., 2009, 'Key website factors in e-business strategy' *International Journal of Information Management* 29, 362–371.

16. Huizingh, E., 2000, 'The content and design of websites: an empirical study', *Information & Management* 37, 123–134.

17. Hussainey, K. & Mouselli, S., 2010, 'Disclosure quality and stock returns in the UK', *Journal of Applied Accounting Research*11(2), 154–174.

18. Kollman, T. & Kuckertz, A., 2010, 'Evaluation uncertainty of venture capitalists' investment criteria', *Journal of Business Research* 63, 741–747.

19. Krauss, S.E., 2005, 'Research paradigms and meaning: A primer', *The Qualitative Report* 10(4), 758–770.

20. Liao, H., Proctor, R. W. & Salvendy, G., 2008, 'Content preparation for cross-cultural e-commerce: a review and a model',*Behaviour and Information Technology* 27(1), 43–61.

21. Liebowitz, J., Rubenstein-Montano, B., McCaw, D., Buchwalter, J. & Browning, C., 2000, 'The knowledge audit', *Knowledge and process management* 7(1), 3–10. http://dx.doi.org/10.1002/(SICI)1099-1441(200001/03)7:1%3C3::AID-KPM72%3E3.0.CO;2-0

22. Nacar, R. & Burnaz, S., 2011, 'A cultural content analysis of multinational companies' websites', *Qualitative market research: An international Journal* 14(3), 274–288.

23. Moustakis, V., Tsironis, L. & Litos, C., 2006, 'A model of website quality assessment', *The Quality Management Journal*13(2), 22–37.

24. Mpindiwa, L., 2010, 'Impact of intranet as a knowledge sharing tool: experiences of ZIMRA', *World Library and Infromation Congress: 76th IFLA General Conference and Assembly*, Gothernburg, Sweden.

25. Panopoulou, E., Tambouris, E. & Tarabanis, K., 2008, 'A framework for evaluating websites of public authorities', *Aslib Proceedings* 60(5), 517–546.

26. Porter, M., 2001, 'Strategy and the internet', *Harvard Business Review*, March, 1–20.

27. Rocha, A., 2012, 'Framework for a global quality evaluation of a website', *Online Information Review* 36(3), 374–382.

28. Saldaña, J., 2009, *The coding manual for qualitative researchers*, Sage, London.

29. Schmidt, S., Cantallops, A.S. & Dos Santos, C.P., 2008, 'The characteristics of hotel websites and their implications for website effectiveness', *International Journal of Hospitality Management* 27, 2008, 504–516.

30. Wacker, J.G., 2008, 'Tragic conceptual flaws of theory – building empirical research', *Decision Line*, October, 10–13.

Chapter 8

Framework to Manage Information Security for Malaysian Academic Environment

Zuraini Ismail[1], Maslin Masrom[2], Zailani Mohamad Sidek[1], and Dayang Suhana Hamzah[1]

[1]Advanced Informatics School, University Technology Malaysia, Malaysia

[2]Razak School of Engineering and Advanced Technology, University Technology Malaysia, Malaysia

ABSTRACT

Among the ICT challenges of most organizations including those from the academic sector is securing their information. Apart from technological aspect, Higher Education Institutions (HEIs) also must implement and enforce proper policies, procedures, and standards in compliance with laws and regulations to safeguard and secure its assets. In this study, four renowned Malaysian HEIs were the choosen in answering the research questions. Two modes of data collection were adopted in this study; interview and survey. Five major constructs of information security framework (ISF) unique to HEIs were identified during preliminary investigation. The result provides valuable information on ISF practices in establishing the components of the ISF. A proposed ISF specific for HEIs was designed. Then a survey was conducted to investigate IT personnel perceptions on the existing information security policy practices at the various HEIs. The result of the study contributes to the understanding of the status and HEIs practices of information security in Malaysian academic setting.

Keywords: Information Security, Information Security Framework, Standards and Policy, Higher Education Institutions

INTRODUCTION

The demography within Malaysian Higher Education Institutions (HEIs) took an evolutionary change since the introduction of Information Technology (IT) in the education environment. The significant achievement of IT in Malaysia can be traced from the early nineties after various adjustments of regulatory and commercial policies, both macroeconomic and within IT's converging sectors (Hancock, 2000). In pace with such adaptation, Malaysia's HEIs are increasingly utilizing IT in all aspects of its organizational functions.

Apart from obtaining benefits from the use of IT, HEIs are also faced with various emerging network security threat that is the result of increasingly sophisticated methods of attack and the blending of once distinct types of attack into more complex and damaging forms (Garuba, et al., 2008). As concluded by McKissack et al. (2010) a gap was noted in terms of the insufficiency of current "best practices where assurance provides confidence on security threats. The survey on the attacks and security incidents reported by MyCERT, CyberSecurity Malaysia for the first quarter (Q1) 2008 revealed that a total of 10,354 security incidents inclusive of spam incidents were reported (MyCERT, 2008). This represented an increase of 5.59% incidents rate compared to fourth quarter (Q4) in 2007 with total of 9,486 incidents. The categories of incidents identified are intrusion, hack threat, malicious code, denial of service and spam. Thus, HEI IT departments in particular must balance between enabling a highly collaborative, non-restrictive environment without discounting confidentiality, integrity, and availability of data and computing resources.

To the best of the researchers' knowledge, empirical research on the enforcement in the context of information security continues to remain an under-investigated research area especially in Malaysian HEIs environment. Hence, this research seeks to answer two research questions, namely

- What are the main components that are considered in the proposed ISF for higher education institutions?
- What are the IT personnel perceptions on the existing Information Security policy practices?

This paper is organized into six sections. This section introduces the study concerned and provides the research questions. Section two portrays the literature review of existing security frameworks. The research methodology is then discussed in section three. Section four narrates the results. The next section explains the survey results. Finally the paper ends with discussion and conclusion.

LITERATURE REVIEW

This section, firstly defines information security. Secondly, the various information security frameworks are examined and then discuss the four standard framework information security commonly used.

Information Security

Information security (InfoSec) is the safeguarding of information, which has a recognized value to any organization including HEIs. It includes both business and technology related aspects. The purpose of information security is to preserve the three elements: confidentiality, integrity and availability (Kasmiran, 2008), in addition, authenticity, accountability, non-repudiation and reliability.

As the business dependence on information technology has evolved so too has the imperative and scope of information security. This emergence has been characterized by Von Solms (2006) in term of four waves: technical, management, institutional and governance. The development of security standards and frameworks coincide with these four waves.

AS/NZ ISO/IEC 27002:206 lists the following success factors as critical to information security:

- The creation of a security framework that is consistent with the organizational culture.
- Visible management commitment to security.
- Provision of security awareness, training and education is not only limited to stakeholders but also to all employees and where relevant to contractors and third party with relevant to their job functions.

- Implementation of a measurement system to evaluate the effectiveness of the security program and provide feedback for improvement of security framework.

Information Security Policy Framework

Framework is define as an the outline action of the more thorough blueprint, which sets out the model to be followed in the creation of the design, selection and initial and ongoing implementation of all subsequent security control. It is also includes information security policies and procedures, security education training programs, and technological controls (Whitman and Mattord, 2007).

From the Information Security (InfoSec) policy perspective, a framework offers a possible starting point for understanding a security policy's impact to an organization, and is intended to guide organizations in developing, implementing, and maintaining security policy (Kasmiran, 2008). Policy should address both logical and physical security. In addition, privacy and confidentiality, integrity and availability, and legal compliance requirements (Computer Associates International Inc, 2005) also should be included.

A primary objective of InfoSec Policy is to define the user's rights and responsibilities in an organization and the effective InfoSec Policy will helps the users understand what acceptable and responsible behavior is in regards to information resources to ensure the safe environment (Hone and Eloff, 2002).

InfoSec Policy has attained an international awareness and several international standards have been built (Hong et al., 2006). The following section will explain the commonly used standards of Information Security Policy.

Existing Information Security Standards Framework

In this section, we discuss the existing standard information security frameworks such as MyMIS, ISO/IEC 27001, COBIT and COSO. Due to the lack of information security framework specifically addressing higher education institutions, the four established framework (MyMIS, ISO/IEC 27001, COBIT and COSO) provided some guideline in developing the proposed information security framework.

MyMIS

Malaysian Administrative Modernization and Management Planning Unit or MAMPU has introduced a handbook called MyMIS. MyMIS basically provides a standard guideline especially for government sector. It comprises of management safeguards, basic operation, technical operation and legal matters (MAMPU, 2002).

ISO/IEC 27001

ISO/IEC 27001 is an information security standard published by the International Organization for Standardization (ISO) and the International Electrotechnical Commission (IEC) as ISO/IEC 17799:2005. It was subsequently renumbered ISO/IEC 27001 and entitled as *Information Technology – Security Techniques – Codes of Practice for Information Security Management*. It is stated that the objective is to serve as a single reference point for identifying the range of controls needed for most situations where information systems are used (IEC 27001, 2005).

Various countries like Australia and New Zealand follows the standard as the basis of regional information security-related standards. ISO/IEC 27001 and its variants provide point description of what should be included as a minimum requirement in information security policy. ISO/IEC 27001 also includes a section exclusively on the review and evaluation of an information security policy.

Among the component outlined by ISO/IEC 27001 are security policy, organization of information security, asset management, human resources, physical and environmental security, communication and operation management, access control, information system acquisition, development and maintenance, information security incident management, business continuity management and compliance.

COBIT

The Information Systems Audit and Control Association & Foundation (ISACAF) developed Control Objectives for Information and related Technology (COBIT) to provide management and business process owners with an IT governance model to help understand and manage the risks associated with IT.

COBIT describes the processes and controls needed for implementing an information security policy, rather than focusing on the document itself. It contains a brief section on the Security and Internal Control Framework Policy, which gives various pointers on writing and maintaining such a document. COBIT consists of four main components namely, plan and organize, acquire and implement, deliver and support, and finally monitor and evaluate.

COSO

Committee of Sponsoring Organizations of the Treadway Commission (COSO) concepts involves organizational internal controls, and is not specific to information security management. However, some aspects of the COSO framework could be adopted for managing security in the enterprise. COSO specially is a business-oriented framework for implementing enterprise risk management (Ritchie and Brindley, 2001).

Based on these known information on existing frameworks, their main components were noted. The next section sets the methodology applied in order to answer the research questions.

METHODOLOGY

In addressing the research questions, two phases of data collection were involved, namely, interview and survey. The interview is to determine the main component of ISF for HEIs. Having completed the interviews, a survey was embarked on in examining the IT personnel perceptions on the existing information security policy practices at respective HEIs.

Interview

The interview conducted involved four (4) HEIs selected located in the Klang Valley. HEIs chosen are all renowned public universities located in the most populous area of Malaysia's economic pulse. For anonymity and confidentiallity reasons, the selected universities are referred as Alpha, Beta, Chi, and Delta.Alpha is a university aspired from a contemporary global Muslim community. Beta University prides to be one of Malaysia's first and oldest universities. Chi University is the only university that has branch campus in every thirteen states of Malaysia.

Delta University is a prestigious HEI set among the top universities in Asia. The university remains to be the main contributor in churning technical expertise. The interviewees comprises of IT-expert staff and personnel in-charge of developing the Information Security framework for each HEI. It was noted that most of the respondents are above 40 years of age; while the number of years in their respective organization and are in the current position recorded as more than five years. This implies that the interviewees are best candidates to provide information on an aggregated unit of analysis in relation to views of information security and its implementation and best practices. Each of the interview exercises lasted from 45 minutes to an hour. It was conducted between the months of May to June of 2009.

The Survey Instrument

The outcome of the interviews contributed towards the design of the survey instrument. The aim of this questionnaire is to investigate IT personnel perceptions on the practices of existing information security policy at HELs. Prior to the actual survey, a pilot study was conducted in order to check the reliability and validity of the instrument constructed. The design of the questionnaire proved to be acceptable. No major alterations were needed.

The questionnaire was divided into six sections. Section one highlights user's profile. Section two captures the risk management in the organization. Section three describes information security policy status. Section four consists of survey on awareness program and training that has been carried out at the organization. Section five defines the access control, followed by compliance concerns in section six.

RESULTS

Through the interviews, Table 1 summarizes the main components considered by the four HEIs in this research.

As depicted in Table 1, various components were included by the respective HEIs. Beta University adopted all eleven domain of ISO 27001. Unlike Alpha considered only five incorporated also those from COBIT. MyMIS is the standard adopted by Chi and Delta while Chi includes technical operations.

Besides this, it is noteworthy to understand the current status of InfoSec policy practices.

Table 1: Components of Main Components Considered by HEIs.

HEI	Component Considered	Standard adopted
Alpha	Risk Assessment Physical and Environmental Security Access Control Information Security Incident Management Compliance	COBIT & ISO 27001
Beta	Risk Assessment Security Policy Organization of Information Security Asset Management Human Resource Security Physical and Environmental Security Communication and Operation Management Access Control Information Systems acquisition, development and maintenance Information Security Incident Management Compliance	ISO 27001
Chi	Management Safeguards Basic Operations Technical Operations	MyMIS
Delta	Management Safeguards Basic Operations	MyMIS

Table 2: Comparison on the Status of Information Security Policy Practices

HLI	ISF	Policies IT	Guidelines IT
Alpha	None	Available	Available
Beta	None	None	Available
Chi	None	None	Available
Delta	None	Available	Available

From Table 2, it is found that none of the HEI has ISF in placed. However, it is minuted that Alpha and Delta both have IT policies in place in comparison to Beta and Chi. Nevertheless, IT guidelines are in place for all the chosen public HEIs.

Hence, this research has summarized the current InfoSec policy practices. With this and the component considered from Table 1, we come up with a proposed conceptual framework suitable to the Malaysian HEIs environment.

Conceptual Framework

The ISF constructs identified for HEIs was adopted from ISO 27001, MyMIS and COBIT. With the established guidelines, we developed the conceptual information security framework as illustrated in Figure 1.

Based on the four (4) interviews, five main constructs were identified to be considered in the HEI ISF. They are information security policy, risk management, access control, awareness program and training, and compliance. Figure 1 shows the conceptual components with their constructs considered for the proposed ISF.

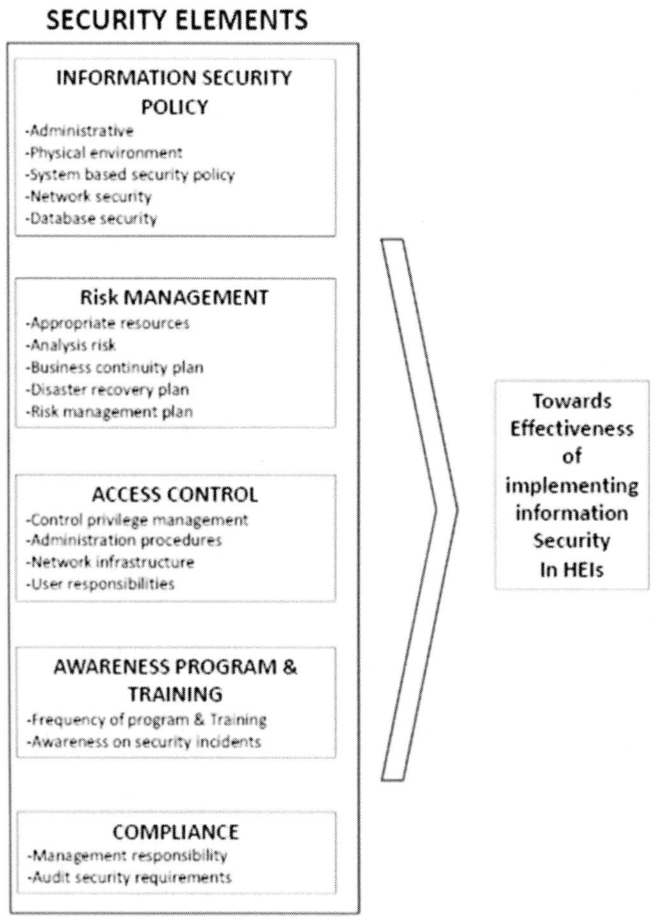

Figure 1. Proposed Conceptual Information Security Framework for HEI

Five security elements are used to enforce the ISF. The security constructs are briefly described below:

- **Information Security Policy** – This part explains about security policies. Security policies control address management support, commitment and direction in accomplishing information security goals including information security policy (Carlson, 2001).

- **Risk Management** – Risk management is the identification and analysis of relevant risks to achievement of the objectives, forming a basis for determining how the risks should be managed (Radack, 2004)

- **Access Control** – Access control addresses an organization's ability to control access to assets based on business and security requirements including business requirement, user management, user responsibilities, network access control, host access control, application access control, access monitoring and mobile computing (Carlson, 2001).

- **Awareness Program & Training** – Based on the awareness on the security issues, and is the single most effective means of ensuring information security. The most effective measures depend largely on the behavior of the people affected by those measures. For example, an access control system based on secret password is effective only if people do not share their password (Elliot et al., 1991).

- **Compliance** - Compliance control addresses an organization ability to remain in compliance with regulatory, statutory and security requirement including legal, technical and system requirements and audits (Carlson, 2001).

Nevertheless, from the four established framework (MyMIS, ISO 27001, COBIT, and COSO) indeed provided beneficial guidelines to the HEIs which in turn form the basis in developing the proposed information security framework depicted in Figure 1.

Survey Results

The survey was conducted in the month of November and December in 2009. The same four (4) HEIs were involved in the data collection in order to address our second research question. With the assistance of human resource officer of the respective HEI's IT department, we identify the possible respondents for this survey. Two hundred (200) were selected and questionnaires were distributed. Seventy-two (72) questionnaires were returned. This interprets to 36% response rate. All returned questionnaires were found to be usable for further analysis.

Respondents Profile

Table 3 portrays the demographic profile of the respondents by gender, age, position in the institution and their years of experience.

Table 3: Respondents Profile

PROFILE		FREQUENCY (N)	PERCENTAGE (%)
Gender	Male	39	54.2
	Female	33	45.8
Age	20-25 years old	18	25.0
	26-30 years old	16	22.2
	31-35 years old	16	22.2
	36-40 years old	7	9.7
	41-46 years old	6	8.3
	More than 46 years old	9	12.5
Position	Project Manager	1	1.4
	IT Executive	14	19.4
	IT Management	7	9.7
	Programmer	36	50.0
	System Analyst	11	15.3
	Others	3	4.2
Year of Experience	1-3 years	17	23.6
	4-8 years	28	38.9
	9-10 years	5	6.9
	More than 10 years	22	30.6
Involved in ISF	Yes	51	70.8

Note: N = 72

As depicted in Table 3 above, slightly over than half of respondents (54.2%) are male. Nearly half of them (47.2%) are below 31 years old and hold the position as programmers. It is noted that more than a quarter of them (38.9%) have between 4 to 8 years working experience in their organization.

In addition, it was found that nearly three-quarters (70.8%) of the participants are involved in the development of Information Security Policy in their organization.

Information Security Policy

This section deals with the status of information security policy available at the respondent's organization. It was found that more than half (56.9%) of respondents are aware of the existence of information security policy. Based on these responses, further explanations are provided.

In response, it was indicated that the policy is consistent, easy to understand, and readily available to administrator (27.29%), faculties (24.47%), staff (24.47%) and student (23.77%). A high majority (95.25%) of the respondents agreed that the policy is reviewed and approved by the top management. In addition, more than three-quarter of the respondents (92.64%) rated 'Strongly Agree' and 'Agree' that the policy was reviewed regularly based on significant changes.

Similarly, most respondents (90.18%) strongly agree that the policy does effectively address the risks identified within their organizations.

Table 4: Awareness of information security policy

No	Questions	Responses			
		Yes		No	
		N	%	N	%
1.0	Is there any indicator to reflect the adequacy and effectiveness of the information security policy?	33	80.49	8	19.51
2.0	Does the information security framework cover the following elements?				
	a) Risk Management	29	70.83	12	5.87
	b) Information Security Policy	40	97.72	1	2.28
	c) Access Control	41	100		
	d) Awareness Program & Training	36	87.87	5	12.13
	e) Compliance	33	80.49	8	19.51
3.0	The areas that the policy has cover				
	a) Administrative	30	73.29	11	26.71
	b) Physical environment	31	75.75	10	24.25
	c) System based security policy	35	85.41	6	14.59
	d) Network security	39	95.25	2	4.75
	e) Database security	33	80.49	8	19.51

Note: N = 41

Question 1.0 of Table 4, it denotes that more than three-quarter of the respondents (80.49%) agreed the policy is effective. The study did reveal that their organization's policy paid attention to administrative (73.29%), physical environment (75.75%), system-based security policy (85.41%), network security (92.25%) and database security (80.49%) area (question 3.0).

There are five (5) constructs listed according to agreement in the information security framework for HEIs. In reference to question 2.0, the result reported that the respondents totally rated to access control (100%) as one of the security elements required for the framework. Subsequently, the respondents rated information security policy (97.72%), awareness program and training (87.87%), compliance (80.49%) and risk management (70.83%) accordingly.

Risk Management

The risk management section seeks to investigate the status of risk management in the organization.

More than half of the respondents (52.8%) are aware of the existence of documented risk management policy in their organization. The survey had showed that scenario analysis is the most preferred technique used to identify risk bearing the highest choice (33.58%) in comparison to the other three techniques. Brainstorming and interview resulted in 27.33% and 23.45% respectively. The respondents identified risk management using survey questionnaire is the least used technique (15.64%).

Table 5: Assessment on Risk Management

No	Questions	Responses			
		Yes		No	
		N	%	N	%
1.0	Does your organization is able to allocate appropriate resources to support current risk management policy and practice?	47	65.3	25	34.7
2.0	Does your organization's response to analyze risks includes :				
	a) An evaluation of effectiveness of existing controls and risk management responses.	47	65.3	25	34.7
	b) An assessment of the cost and benefits of addressing risk.	58	80.6	14	19.4
	c) Prioritizing of risks and selecting those that need active management.	57	79.2	15	20.8
3.0	Does your organization have the following security measure documents?				
	a) Business continuity plan	42	58.3	30	41.7
	b) Disaster recovery plan for information technology.	64	88.9	8	11.1
	c) Risk management plan.	52	72.2	20	27.8

Note: N=72

More than half of respondents (66.7%) had agreed that it is very important to have effective risk management in achieving organization's objectives. More than quarter (27.8%) indicated that it is somewhat important and a few (5.6%) thought that it is not important at all. On the other hand, a majority (95.8%) of the respondents had rated 'Strongly Agree' and 'Agree' on the issue that effective risk management can improve organization's performance.

From the Table 5 above, question 1.0 reads slightly more than half (65.3%) of the respondents clearly defined that their organizations is able to allocate appropriate resource to support current risk. Question 2.0 portrayed that more than half of the respondents agreed that their organization give response to analyze risk including the evaluation of effectiveness (65.3%), assessment of the cost (80.6%) and prioritizing of risk (79.2%). More than half respondents also stated that their organizations has the following security measure documents which are Business Continuity Plan (58.3%), Disaster Recovery Plan (88.9%) and Risk Management Plan (72.2%). The result is shown in question 3.0.

Awareness Program & Training

In this section, result of status awareness program and training conducted in the organization will be present. Table 6 shows percentage of security awareness program and training conducted for administrator, faculties, staff and students (question 1.0). More than half of the respondents stated the training are frequently held for administrator (73.6%) and staff (75.0%). As for the faculties (58.3%) and student (40.3%) seems to record lower agreement on security awareness training.

In examining the duration of awareness program and training, nearly a third of respondents (30.6%) stated that it was conducted within six months to one year. About another third (31.9%) reveals that the program duration is more than one year. It is also reported that slightly over a third of the respondents (37.5%) stated that the training is conducted less than six months. It is noted that more than half of the respondents (52.8%) stated that their training program was conducted by both internally and other external bodies, with nearly half of respondents (44.4%) reported that it was conducted internally. The rest indicated that their training was conducted by other companies.

Table 6: Determining the awareness program & training

No	Questions	Answer			
		Yes		No	
		N	%	N	%
1.0	Does your organization conduct any security awareness program & training for				
	a) Administrator	53	73.6	19	26.4
	b) Faculties	42	58.3	30	41.7
	c) Staff	54	75.0	18	25.0
	d) Student	29	40.3	43	59.7
2.0	Are you aware of your organizations or departments ICT security policies, procedure and guidelines to protect information?				
	a) Security policies	60	83.3	12	16.7
	b) Procedure	56	77.8	16	22.2
	c) Guidelines	60	83.3	12	16.7

Note: N = 72

In the case of computer security incident occurrence, most of respondents (84.7%) prefer to contact the person in-charge to fix the problem. Very few (8.3%) would prefer to self-fix the problem. Very fewer still (6.9%) choose to report the problem to administrator via e-mail.

Access Control

This section will present a result of access control status at the respondents' organizations. The result is shown in Table 7 below.

Table 7 reveals nearly three-quarter (73.6%) of the respondents are aware of the existence of written policy for the use network service in their organization (question 1.0).

Most of them are also aware of the existence of access control procedure (79.2%) and policies (75.0%) as to support the access control policy. This has shown in question 2.0. Based on question 3.0, more than three-quarter of the respondents (81.9%) stated that their organization have control privilege management. The result reported that their organization has protect secured areas by appropriate entry control (95.8%) (question 4.0), do a check on equipment containing storage media (80.6%) (question 5.0) and separate the development, test and operational facilities to reduce risk (79.2%) (question 6.0). Based on question 7.0, respondents agree that both network infrastructure (88.9%) and administrative procedures (75.0%) are maintained and updated.

A high majority of respondents (94.4%) stated that their organization use a Virtual Private Network (VPN) (question 8.0). More than three-quarter (98.6%) of the respondents reported that all connection with external network is protected by firewall (question 9.0). According to question 10.0, more than three-quarter of the respondents (87.5%) reported the firewall configurations regularly reviewed and updated. It is further noted, more than three-quarter of respondents defined the way they monitor their network traffic by bandwidth statistic (91.7%), top protocol bandwidth consumer (87.5%) and top IP bandwidth consumer (84.7%) (question 11.0).

Table 7: Access control concerns

No	Questions	Answer			
		Yes		No	
		N	%	N	%
1.0	Has management developed and published a written policy on the use of network services?	53	73.6	19	26.4
2.0	Do access control procedures and policies exist to support the access control policy? a) Access control procedures b) Policies	 57 54	 79.2 75.0	 15 18	 20.8 25.0
3.0	Does your organization have control privilege management for information systems and applications?	59	81.9	13	18.1
4.0	Are secured areas protected by appropriate entry controls to ensure that only authorized personnel can use the facilities?	69	95.8	3	4.2
5.0	Are all items of equipment containing storage media checked to ensure that any sensitive data and licensed software has been removed or securely overwritten prior to disposal?	58	80.6	14	19.4
6.0	Are development, test and operational facilities separated to reduce the risk of unauthorized access or changes to the operational system?	57	79.2	15	20.8
7.0	Are the network infrastructure and administration procedures updated and maintained? a) Network infrastructure b) Administration procedures	 64 54	 88.9 75.0	 8 18	 11.1 25.0
8.0	Does your organization use a Virtual Private Network (VPN)?	68	94.4	4	5.6
9.0	Are all connections with external network protected by firewall?	71	98.6	1	1.4
10.0	Are the firewall configurations regularly reviewed and updated?	63	87.5	9	12.5
11.0	How do you monitor your network traffic? a) Bandwidth statistic b) Top protocol bandwidth consumers c) Top IP bandwidth consumers	 66 63 61	 91.7 87.5 84.7	 6 9 11	 8.3 12.5 15.3

Note: N = 72

COMPLIANCE

Compliance section seeks to examine the status of compliance in the organization. The result is presented in Table 8.

Table 8: Assessment on Compliance

No	Questions	Answer			
		Yes		No	
		N	%	N	%
1.0	Is there a person or committee that has information security as their primary duty?	62	86.1	10	13.9
2.0	Does your information security policy have the authority it needs to manage and ensure compliance with the information security policy?	56	77.8	16	22.2
3.0	Is someone in the information security policy responsible for liaising with units to identify any new security requirements?	63	87.5	9	12.5
4.0	Does the information security policy actively engage with other units (human resources, student affairs, faculty and library) to develop and enforce compliance with information security policies and practices?	52	72.2	20	27.8
5.0	Does the information security department report regularly to the governing board on the compliance and the effectiveness of the information security program and policies?	47	65.3	25	34.7
6.0	Are there specific training programs in place to comply with the information security and standards with the goal of ensuring the security of the information systems that support the operations and assets under control?	45	62.5	27	37.5

Note: N = 72

Based on question 1.0 in the table above, more than three-quarter of respondents (86.1%) indicate that there is a person or committee in-charge in securing their organization's information. In question 3.0, slightly more than three-quarter (77.8%) of the respondents stated that there is also a person in-charge to liaise with other units to identify any new security requirements to the policy.

Slightly more than three-quarter of respondents (77.8%) agreed that their organizations' information security policy has the authority to ensure compliance with the policy (question 2.0). Still, a majority (87.5%) of the respondents agreed that the policy is actively engage with other units to enforce compliance with information security policy and practices (question 4.0).

As to ensure the effectiveness and compliance of information security program and policies, two-third of the respondents (65.3%) indicated that the information security department reports regularly to the governing board (question 5.0). Furthermore, result of question 6.0 showed that more than half of the respondents (62.5%) agreed that there are specific training programs to comply the policy and standards are necessary.

DISCUSSION AND CONCLUSION

This study has successfully answered both the research questions. Firstly, the main constructs of HEIs ISF were determined through interviews. Secondly, based on the IT personal responses through a survey, it establishes their perceptions on the existing information security policy practices.

Through the interviews, it revealed that largely all the established government-supported HEIs have some form of IT policies in place. Various establish information security standards were found to form the basis of their efforts towards the development of their very own policy.

This research further discovers that information security policy, risk management, access control, awareness program and training, and compliance are the major components or elements that fix upon the suggested framework for HEIs. These components are in line with the objective of developing the framework which was to apply and cover all hardware, software, data, information, network, personal computing devices, support personnel, and users within HEIs from intrusion, interception, interruption and denial of services.

Through the survey, generally, the respondents who largely were involved in the development of their organization's InfoSec policy perceived the importance of each component of the proposed framework.

It was found that using a scenario analysis as a technique to identify risk is found to be more effective since the technique is designed to allow improved the decision-making. It considers the outcomes and their implications. In order to have a good risk management policy, organization also must be able to allocate an appropriate resource to support the existing risk management policy. The risk analysis should further emphasize on the assessment of the cost and benefits of addressing risk, similarly prioritizing of risks and its impact and likelihood besides selecting those that need active management and evaluation on the effectiveness of existing controls and risk management responses. Furthermore, most of the organizations have realized that having the disaster recovery plan, risk management plan, and business continuity plan are important in order to protect their organizations data and IT infrastructure in the event of the disruptive situations. This research also found that having an effective risk management can improves organizations' performance and simultaneously, it also helps towards achieving the organizations' objectives.

This study does disclose top management concerns on the importance of adequate information security policy in the organizations. This is indicated by periodically reviewing and updating the policy based on significant changes due in relation to the risk identified by the organizations. Thus, the policy is consistently and readily made available for compliance by the administrator, faculties, staff, students and third party. Accordingly, the findings indicated the effectiveness of the policy the organizations have addressed all the five security constructs identified in the ISF. Access control is found to be the most important security element. Additionally, policy also plays an important role in explaining to staff and students of their responsibility in the protection of the information resources, while stressing the importance of having secured information.

Awareness and compliance is the success key to the policy as implementation will take place after the policy had been endorsed. It was showed that the awareness training has been conducted to administrator and staff in the organization more frequent than for faculties and the students. As the students form the majority of the campus population does it remains imperative to suggest more awareness program towards information security exposure. However, the study showed no bearing on the duration of the awareness program and training in relationship to awareness acceptance specifically.

This study also indicated that HEIs appoints specific person in-charge or committee set up in addressing ISF and policy concerns. In furtherance of ensuring the compliance enforcement, information security policy department is actively in cooperation with other units of the organization and regular meetings and scheduled reporting are practiced.

Hence, this effort would provide some clarification and insights into how ISF is depicted in the academic setting. Further work is obviously necessary to look into the details of the framework. Amongst others, matters pertaining to third party, asset management, equipment security, communications, systems acceptance, cryptography and incident management. Thus, this research serves as an expansion of security and assurance in operational areas literature in the area of ISF engagements.

Despite the study's limitations, we believe that our work makes significant contributions to practice and research.

1. Managerial perspective

a) It provides as an indicator to the status of ISF implementation from IT personnel perspective.

b) It identified the various standard adopted by HEI or the lack of it.

c) HEIs can enjoy significant benefits from making right choices in terms of construct that are relevant or set a priority- level to its ICT-related activities.

d) Serves as preparatory guidelines for future planning and improvements to HEIs ISF.

e) Better understanding of critical ISF construct to ensure successful enforcement of information security.

2. Theoretical Contribution

a) Clarification and rearrangement of the available constructs: information security policy, risk management, access control, awareness program and training, and compliance are delineated in the proposed ISF model

b) Identification of new constructs whereby the involvement of top management plays a significant role in sustaining a robust and effective. Notably, adequate implementation and improvement of the ISF hinges on management's commitment.

In a nutshell, this study forms a basis in understanding the status and its practices of information security in Malaysian academic setting. This will further fulfill the HEIs information security needs towards a more dynamic yet sustaining a secured academic environment.

ACKNOWLEDGEMENT

This research is funded by e-Science grant vote number 79317 awarded by the Ministry of Science and Technology Innovation (MOSTI) of

Malaysia. Our sincere thanks go to Malaysia's Ministry of Higher Education (MOHE), and the cooperation of participating universities.

REFERENCES

1. AS/NZ ISO/IEC 27002:2006 (2006), 'Information Technology – Security Techniques – Code of Practice for Information Security Management,' *AS/NZ ISO/IEC 27002.*

2. Computer Associates International, Inc. (2005). 'Best Practices: Security Management.' *White Paper published in Partnership Inc.*

3. Elliot, R., Young, O.M., Collins, D.V., Frawley, D. and Temares, L.M. (1991),'Information Security In Higher Education,' *Cause*, 1-35.

4. Garuba, M., Liu, C. and Washington, N. (2008). 'A Comparative Analysis of Anti-Malware Software, Patch Management, and Host-Based Firewalls in Preventing Malware Infections on Client Computers' Proceedings of the Fifth International Conference on Information Technology: New Generations, IEEE Computer Society, Washington, 628-632.

5. Hancock, B. (2000), 'New York Times Fired Employees for Violating Internal E-mail Policies,' *Computer Fraud and Security*, 12, 20.

6. Hone. K. and Eloff. J.H.P. (2002), 'What Makes An Effective Information Security Policy?' *Network Security,* 6, 14-16.

7. Hong, K.-S., Chi Y.-P., Chao, R. L., and Tang J. H. (2006), 'An Empirical Study Of Information Security Policy on Information Security Elevation in Taiwan,'*Information Management & Computer Security.* 14 (2), 104-115.

8. Kasmiran, J., (2008), 'The Security Factor In The Information Technology', [Online], [Retrieved February 14, 2009],http://pkukmweb.ukm.my/pkukm/content/view/84/94/

9. MAMPU --- Malaysian Administrative Modernization and Management Planning Unit, (2002), MyMIS - *Malaysian Public Sector Management of Information & Communication Technology Security Handbook,* Putra Jaya.

10. McKossack, J., Hooper, V. and Hope, B. (2010), 'An Orgdel for Information security Assessment,' Proceedings of the International

Conference on Information Management and Evaluation(ICIME 2010), ISBN: 978-1-906638-56-6, 25-26 March2010,Reading, England, 218-227.

11. [Online] [Retrieved January 22, 2009] http://en.wikipedia.org/wiki/ISO/IEC_27001

12. [Online] [Retrieved February 5, 2009]http://www.mycert.org.my/en/services/advisories

13. /mycert/2008/main/detail/579/index.html

14. Radack, S. (2004) 'Risk management framework: Helping organizations implement effective information security program,' *INS Whitepaper*, Santa Clara, INS.

15. Ritchie, B. and Brindley, C. (2001), 'The information-risk conundrum,' *Marketing Intelligence and Planning,* 19(1), 29-37.

16. Carlson, T. (2001), "Information Security Management: Understanding ISO 17799",*INS Whitepaper*, Santa Clara, INS.

17. Von Solms, B.,(2006), 'Information Security – The Fourth Wave,' *Computer and Security*, 25, 165-168.

18. Whitman, E.M. and Mattord, J.H. (2007), 'Management of Information Security,'Thomson Course Technology, Boston.

Chapter 9

Comparative Case Study on Website Traffic Generated by Search Engine Optimisation and a Pay-Per-Click Campaign, Versus Marketing Expenditure

Wouter T. Kritzinger, Melius Weideman

[1]*Website Attributes Research Centre (WARC), Cape Peninsula University of Technology, South Africa*

ABSTRACT

Background: No empirical work was found on how marketing expenses compare when used solely for either the one or the other of the two main types of search engine marketing.

Objectives: This research set out to determine how the results of the implementation of a pay-per-click campaign compared to those of a search engine optimisation campaign, given the same website and environment. At the same time, the expenses incurred on both these marketing methods were recorded and compared.

Method: The active website of an existing, successful e-commerce concern was used as platform. The company had been using pay-per-click only for a period, whilst traffic was monitored. This system was decommissioned on a particular date and time, and an alternative search engine optimisation system was started at the same time. Again, both traffic and expenses were monitored.

Results: The results indicate that the pay-per-click system did produce favourable results, but on the condition that a monthly fee has to be set aside to guarantee consistent traffic. The implementation of search engine optimisation required a relatively large investment at the outset, but it was once-off. After a drop in traffic owing to crawler visitation delays, the website traffic bypassed the average figure achieved during the pay-per-click period after a little over three months, whilst the expenditure crossed over after just six months.

Conclusion: Whilst considering the specific parameters of this study, an investment in search engine optimisation rather than a pay-per-click campaign appears to produce better results at a lower cost, after a given period of time.

INTRODUCTION

The growth of the Internet has produced an important information resource during the last two decades that has advanced at a much faster rate than was previously envisaged. It took seven years to reach a 25% international market share – 70% faster than the development of the radio and 80% faster than the development of the telephone. This growth makes the Internet the fastest growing technology the world has ever encountered (Singh 2002). Boyes and Irani (2004:191) support this trend by claiming that the Internet had acquired 50 million global users in five years as opposed to the 38 years it took for radio and 13 years for television.

The implementation of the World Wide Web (WWW) has seen the world confronted with the concept of a website. Websites act as connection and communication points between the user and digital information. Therefore, most corporations (according to Akakandelwa 2011), organisations or institutions have been making efforts to launch themselves into the virtual world using this modern platform. The WWW is more than two decades old and, due to its complexity, its size is impossible to measure with regard to the number of websites or servers. It is claimed that, for January 2014, the nine most popular websites in the United States of America (USA) drew between 100 million and 370 million visitors each (Nielsen 2014). The WWW is a decentralised environment constructed and controlled by various people and access to it is less restricted than access to the common information media (Brunn & Dodge 2001).

The base of Internet users is massive; hence, there is much interest in leveraging this user base for commercial gain. This commercial gain could be realised by ensuring that many thousands of users view a given website daily, with some of them being converted from browsers to buyers. It is generally accepted that the two types of interventions which could be implemented to increase the traffic to a website are search engine optimisation (SEO) and a paid campaign. However, no research could be found which compares the expenditure in a controlled environment with the value received from that expenditure for these two approaches. It is against this background that this study has originated.

The research problem on which this project was based is the fact that the respective value of the two marketing methods has not been directly compared, leading to resources being wasted on marketing. The purpose of this study was to compare these two categories to see how they produce traffic over a period of time, and to offset them against the expenditure.

LITERATURE REVIEW

Search engines

According to Green (2000), a search engine is a search service that uses retrieval software called crawlers that examine websites and then index them in a database of website listings according to their relevancy. Search engines use their own indexing software and strategies to continuously traverse the Web, searching for the most up-to-date content possible. The indexing software (also referred to as spiders or bots) is responsible for visiting webpages following links between pages. The pages found are then analysed, parts are copied back to the site running the indexing software and added to the database for the purpose of including them in the search engine results (Weideman 2009).

Even though search engines use different algorithms to rank a webpage, they operate on similar principles. All search engines, primarily, strive to retrieve and display relevant results (webpages) that contain words or terms that match the user's search query (Green 2000; Guenther 2004:47).

Oppenheim *et al.* (2000:191) are of the opinion that, although search engines search a vast amount of information at impressive speeds, they are criticised on issues such as the retrieval of duplicate and irrelevant records owing to spamming techniques. The sheer mass of these irrelevant results is one of the main user complaints against search engines.

Much research has been done recently on search engines, the way they operate, the way users generate queries, and the way they affect our lives. Moreno and Martinez (2013) indicated that webpages can be designed to be both SEO and human friendly. This was confirmed by another study on the effect of usability elements on the way search engine crawlers view websites (Visser & Weideman 2011).

In summary, the availability and use of search engines affect our lives on a regular basis through user interaction with results produced by them. It is now certain that the use of search engines, together with the power of advertising, plays a role in peoples' daily decisions on various aspects and is not limited to purchasing only.

Search engine marketing

Search engine marketing (SEM) is a strategy that makes use of the power of search engines to potentially attract millions of views per day to websites. Even academic universities need to consider how they should market themselves to prospective students as they are the paying clients (Weideman 2013). Crowley (2014) discusses some of these strategies, including SEO and tracking analytics, to determine user behaviour. Many systems are fighting for 'eyeballs'; these are users reading and responding to advertisements on websites. Social media has added to the frenzy through their advertisement offerings, although research has proven that it has not been all that successful. Barreto (2013) indicated that Facebook banner advertisements have less business value than recommendations from friends, possibly because of banner blindness. These ads (advertisements) were located outside the main viewing area first scanned by most users.

A recent study was done to determine the effectiveness of a search marketing campaign and the effect of print advertising on SEM (Olbrich & Schultz 2014). It was found that the budget and the degree of keyword matching had the largest effect on the yield in business gained, followed by the click-through rate and the bid amount. This result seems to indicate that SEM needs to be planned with care as a larger budget might reduce expenditure on other marketing efforts. This fact supports the current study. SEO and pay-per-click schemes (PPC) are generally considered to be the two main categories of SEM which are to be investigated.

Pay-per-click

PPC schemes, producing non-natural rankings, are systems which display advertisements on a search result screen, co-located with organic results but ranked separately. The location of these advertisements is normally to the right and above the organic search engine listings (Chen *et al.* 2011). This sharing of the prime real estate space on the user's screen has caused some problems for users. No longer can they simply evaluate the quality of answers based on which answer is listed the highest on the result screen as they have to also consider the ranking difference between organic and paid results. Even libraries have found this to be an obstacle for their users (Moxley, Blake & Maze 2004). However, many industries, particularly the tourism sector, have been making extensive use of this marketing opportunity. In some cases even small businesses went to extremes to do research to identify the better search engine PPC scheme to use (Kennedy & Kennedy 2008). At the same time, the 'newness' of this form of marketing initially scared off other smaller businesses for example some smaller hotels did not make use of PPC (Murphy & Kielgast 2008).

PPC, as the name suggests, charges the advertiser the bid amount every time an Internet user clicks on an advertisement. The keywords all have different competition ratings and the more popular a keyword, the higher the cost per click would be (Chen *et al.* 2011). A recent study on the value of the bid price per keyword for a new PPC campaign determined the best of a number of methods to determine this price (Nabout 2015). This study confirms the importance of financial expenditure, which could easily grow beyond what a company decided they could spend on a marketing campaign.

The PPC ranking system operates in stark contrast to the value associated with the quantity and quality of inlinks, in other words SEO (Thelwall 2001). In the past Google specifically used a simple formula to determine the ranking of PPC results: Rank = Bid price X Quality Score (Sagin 2013). The bid price is the amount the advertisement owner is prepared to pay per user click on the ad, and the quality score is Google's interpretation of the quality of the landing page. However, in October 2013 Google announced a third factor, namely ad extensions (Mancuso 2013). Where two competing advertisements achieve an equal score, the use of ad extensions is used to determine the highest ranker.

A recent study was done on the relationship between print and search engine advertising (Olbrich & Schultz2014). The results proved that print advertising did not directly affect the number of advertisements impressions produced by the search engine. However, there was an indirect relationship between print advertising and the number of conversions indicating that e-commerce marketers cannot ignore the traditional advertising methods and focus only on SEM. Since exposure through PPC requires constant and accumulative expenditure, past research has also focused on maximising this expense. The performance of PPC advertisements is crucial in recapping the expense, hopefully bypassing it to provide a profit. Another recent study determined the role of ranking of these advertisements, branding, and the role of the device used to search (Gupta & Mateen 2014).

In summary, PPC has been a successful business model. In the case of Google, it has produced the bulk of its profits over the years, playing a major role in its financial success (Kumar & Kohli 2007). At the same time, an e-commerce business running PPC schemes on multiple keywords across many campaigns is advised to budget for specialised staff to manage these campaigns. Still, some authors actually prescribe that PPC is a better way to spend marketing dollars than SEO (Sen 2005).

Search engine optimisation

SEO is a method that uses data observation and marketing research to identify the most suitable keyword for a website (Malaga 2010). However, it requires a base of knowledge to implement, such as how to choose keywords and how to use keywords in order to enhance a website's ranking, etc.

There is a need for any e-commerce business to be ranked highly with search engines (Kent 2012). Another author states that to increase the volume of traffic to individual websites from search engines, SEO must be considered and invested in (Kisiel 2010). For successful SEO there are many concepts that need to be understood and applied; some of these are discussed later. The ultimate goal of SEO is to increase a website's ranking with search engines, thus increasing the traffic to the website, which should result in increased sales (Lee 2010; Lee, Chen & Wu 2010; Lee & Lin 2011).

Lately, much research has been done on the visibility of content to search engine crawlers. Onaifo and Rasmussen (2013) found that the principles of SEO can, and should, be applied to increase the visibility of library content to search engines. A number of elements affect SEO; it is not just a case of implementing a simple set of rules, thereby ensuring high visibility. Some of these elements are listed below.

Another recent study considered the value of using search query data to obtain business information. Search query data is much more recent than, for example, business reports published at the end of a financial year. A significant correlation was found between business performance and position and search query data (Vaughan 2014). This kind of timely information could be used to predict business performance, leading to financial gains.

It is clear that both SEO and PPC could play a role in marketing a website to the search engine crawlers. The purpose of this study was to determine how much traffic each of SEO and PPC produces and to measure and compare the expenditure in each case.

Some of the aspects of SEO are considered below. They are all discussed with regard to the components of a webpage over which the website owner has direct control.

Header tags

Metatags are elements of a webpage that are mostly optional, invisible to the casual browser, but which could affect the way a crawler views a webpage. One of these is the H (Heading) meta tag.

A Web designer can choose whether or not to highlight headings inside a block of text on a webpage. To this effect, there are six levels of heading tags. H1 is the most important (biggest text); H2 is slightly less important, down to H6 which is the least important. Some search engines recognise the use of header tags as a safe method to weight keywords, owing to its connection with a heading of a paragraph. Henzinger, Motwani and Silverstein (2002:9) state that the higher the importance of the headings, the more weight a search engine could assign to a given webpage. For example, text in an <H1> would appear prominently on a webpage and therefore some search engines could see it as safe to assign a high weight to the text in that heading.

Research by Craven (2003) to determine the relevant weight of meta tags indicates that the H1 (heading 1) and H2 (heading 2) tags are the second and third most highly weighted (after the TITLE tag) of all the meta tags. As with Microsoft Word, the hypertext mark-up language (HTML) has built-in styles for headings to differentiate between importance levels of text that are usually used to break up text into paragraphs. The different options allow the designer to develop large and bold text in an HTML document, marking the beginning of a new paragraph or section (Henzinger *et al.* 2002:9).

Image filenames

Search engine crawlers cannot interpret the content of an image, a video or an audio file. The designer has to provide text-based information to allow the crawler to make some association between this type of file and its contents (Weideman 2009). For still images, the name of the file is the most obvious way of establishing this association. A simple experiment can prove that the name of an image file does play a role in its visibility. When doing an image search on Google or Bing for the term 'rolls royce' for example, page after page of images of this car are produced. Upon closer inspection of the first result page, virtually every image has those two words as part of the file-name.

Image alternative tags

ALT (alternative) tags are used to display text in the place of an image on a webpage if graphics are turned off. The ALT text will also display if a user places his or her mouse pointer over an image for a few seconds. Currently, automated crawlers can read only text elements within a webpage and are unable to read multimedia elements, as discussed earlier. For this reason, it is of importance to apply ALT tags, where possible, that accurately describe the graphics on the webpage (Hubbard 2004).

By implementing ALT text within a webpage, the developer ultimately caters for users who turn graphics off to increase loading speed. Without the use of ALT tags a site could become impossible to navigate when graphics are turned off. The use of ALT tags also provides the ability to cater for users with a visual impairment. Voice-output screen readers (benefiting those users) will not 'read' a non-text element (Oppenheim *et al.* 2000:204), but will do so if an ALT element is provided. Ironically, the implementation of techniques to allow the visually impaired to better interpret website contents could also play a major role in improving the user experience of other users, as well as the way crawlers interpret webpages.

Metadata

Another meta tag which plays a role in visibility is the TITLE tag. It is also invisible as part of the body text of a webpage being displayed, but it is often displayed on a user screen as part of search engine results. Search engines often claim that the presence of a well-written TITLE tag can positively influence a webpage's visibility (Guenther 2014).

In summary, only a few of the many elements affecting the crawler visibility of a webpage have been discussed. There are many others (keyword usage, HTML file-names, links, etc.), as well as a long list of negative elements that should specifically not be implemented (Weideman 2009).

METHODOLOGY

Data were gathered from a real-life website where first the one (PPC) and then the other (SEO) of the marketing approaches were followed exclusively. Usage behaviour and statistics were recorded and analysed in an attempt to compare the expenditure with the gain from each approach.

A company in Cape Town, South Africa, manufactures PVC, polypropylene and leather promotional and stationery products. The company's name is not listed; for the purposes of recording they will be named XYZ. The company invested in PPC from May 2010 to May 2011 in an attempt to drive traffic to the website. During that time no SEO implementation was done on the website. Each month XYZ spent, on average, R3000 on PPC. However, after the PPC campaign was terminated, they paid a once-off fee of R19 000 for the SEO project.

The XYZ concern is a relatively small company. Considering that they have a typical monthly website traffic figure of hundreds as compared to millions for large companies, their business model must be sound to run a successful e-commerce concern. The most important requirement of the client was that their website remained optimised in such a way that it could lead to sales. The authors did a detailed investigation of the XYZ website, identifying elements that could be improved through SEO with the estimated short-term cost it may entail.

The following elements which needed improvement were identified:

- Header tags (were not being used).
- Image filenames (were inappropriate and non-descriptive).

- Metadata (was outdated).
- Alternative tags (were inappropriate and non-descriptive).
- Product descriptions (were too short or non-existent).
- The bounce rate (was too high).

Header tags

All the headings on the XYZ website were placed in strong tags. Best practice prescribed that all strong tags be replaced with H1 and H2 Tags. This would imply that the authors would have to inspect every one of the 55 webpages and either write new H1 tags, or rewrite the existing ones.

Image filenames

Some of the image filenames did not contain relevant keywords. An example of this was '_MG_2174.jpg'. Again, best practice was that these filenames be renamed with appropriate keywords that reflect the objects on each image. A total of 246 images were found and renamed in the 55 HTML pages.

Metadata

Throughout the year the website was updated with new products, whilst older products were removed. However, the metadata was not updated accordingly. It was recommended that the metadata of all 55 HTML pages be updated to reflect the new content. This was done, and it included updating the TITLE-, DESCRIPTION- and KEYWORD tags of all 55 HTML pages.

Image alternative tags

Some of the ALT tags of the images were too long. These ALT tags, in accordance with best practice, needed to be shortened and rewritten to contain relevant keywords that reflect the images that they are associated with. There were a total of 256 images whose ALT tags were reviewed and rewritten.

Product descriptions

Some of the products listed on the XYZ website were identified by only one short sentence. It was decided, based on best practice, that all product descriptions be rewritten to have at least two to three descriptive sentences each. Sixty four of the 128 products did not have sufficient descriptive text and were subsequently rewritten. The remaining 64 descriptions were also reviewed to ensure that they contained relevant keywords.

Bounce rate

The XYZ website had a bounce rate of 48%, where a lower figure is considered to be better (Plaza 2009). In industry, a bounce rate of 30% is considered to be a breakpoint; higher than 30% is 'bad' and lower is 'good'. It was necessary that the homepage text be reviewed with the aim to reduce the high bounce rate. However, the bounce rate is not an element that a Web developer can improve by doing coding updates or any other direct actions. It is a figure which will decrease (i.e. improve) as users spend more time on webpages as a result of other elements described here being put into place.

Project plan

The project was thus split into three stages:

- Throughout June 2011 the header tags were modified.
- Throughout July 2011 the image filenames and image ALT tags were modified.
- Throughout August 2011 the product descriptions, copy, and metadata were modified.

These timescales appear in the figures following, to indicate the effect of each alteration.

RESULTS AND INTERPRETATION

Header tags

The effect (if any) of the header tag improvement, as discussed earlier, was measured by viewing statistics on website usage as from when the header tags were redone. Refer to Figure 1.

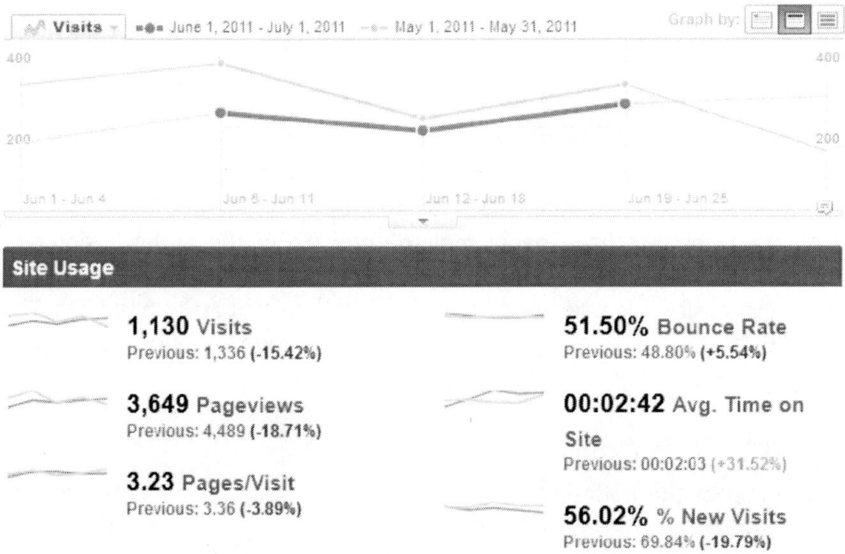

FIGURE 1: Effect of header tag improvement on website visitations

The company stopped their PPC campaign on 31 May 2011. The immediate effect of this action was a drop in total traffic by 15.42% (1130 vs 1336) when comparing June 2011 with May 2011. The authors then proceeded to review and modify all headings on the website. This process involved modifying the CSS file and defining the Header Tags. There were 244 instances where strong tags were used, and all of them were replaced with H1 and H2 tags.

Also, the headings were modified to include appropriate keywords. The authors ensured that each page only had one instance of the H1 tag, which is considered the most important of the heading tags. The work was conducted throughout June 2011 and was completed in the final week of June 2011. It must be noted that all the work was completed on the live website. As the authors progressed a slight increase in traffic in the second week of June 2011 was noted.

It must also be noted that 01 May 2011 and 02 May 2011 were Worker's Day and a public holiday respectively in South Africa. These two calendar events would have had a negative impact on the statistics as the biggest part of the workforce would not be at work on those days. However, this pattern repeated the next month: 16 June 2011 and 17 June 2011 were Youth Day and a school holiday respectively. These two calendar events would have had a similar effect on the figures of June 2011. Also, during 25 June to 18 July 2011 the schools in South Africa were closed for recess which would also have had a negative impact on the June 2011 statistics.

By this time half of the 244 instances of strong tags were replaced by either H1 tags or H2 tags. The fact that the authors saw an increase in organic traffic was somewhat unexpected, since it can take search engines up to a month (Zuze & Weideman 2013) to re-index and rank webpages after significant changes have been implemented. From the author's point of view this highlights the importance search engines place on Header tags. For any website it is a simple matter to assign header tags to headings. The website, from a search engines perspective, was seen in a more favourable position because of the above action taken.

ALT tags

The effect of the updating of the ALT tags was also measured by viewing the relevant statistics – see Figure 2.

The modifications of the image ALT tags were completed in the final week of July 2011. Whilst performing this task the authors observed a slight decrease in traffic from the second week to the third week of July 2011. However, a slight increase in total traffic followed the initial decline as the month of July 2011 drew to a close. By this time, the total number of visitors dropped from 1336 (May 2011) to 825 (July 2011). As mentioned previously, during the month of June the holiday events would have impacted on these figures. Some users at that time would have gone on family holidays, reducing Internet activity to an extent.

It is expected to note decreases in traffic when extensive work is being performed on a website. Search engines take note of content changes after which they have to re-index and rank the individual webpages. This process can take anything from a few days to months. The increase by the end of July 2011 was also, in a sense, unexpected as the authors argued that the process of re-indexing and ranking would take longer than it did. This increase also indicates the value basic SEO elements can have on a website's ranking.

FIGURE 2: Effect of image file-name and image ALT improvement on website visitations.

Body content

The role of body text and the use of keywords have been noted many times in previous research. As mentioned above, sections of this body text had to be rewritten and the graphs of Figure 3 indicate the effect of these changes.

A lack of regular updating of HTML metadata content was noted in the webpages of the company. Whilst descriptive text in the body was updated, the relevant metadata was simply left with the original content, leading to a severance of the connections between these two elements.

Care was taken to ensure that all three meta-elements were rewritten according to Google's guidelines. The new version contained relevant keywords and reflected the product that it related to. This work was completed in the last week of August 2011.

The August 2011 total traffic increased compared to July 2011 (16.36%). This indicated to the authors that Google was starting to re-index the text website with the modifications. It must also be noted that the schools in South Africa were closed for recess between the periods of 24 June 2011 and 18 July 2011, which would have negatively impacted the statistics for the month of July 2011. However, August 2011 did have two public holidays on 08 August 2011 (school holiday) and 09 August 2011 (Women's Day). These two calendar events would have had a similar negative effect on the August 2011 figures.

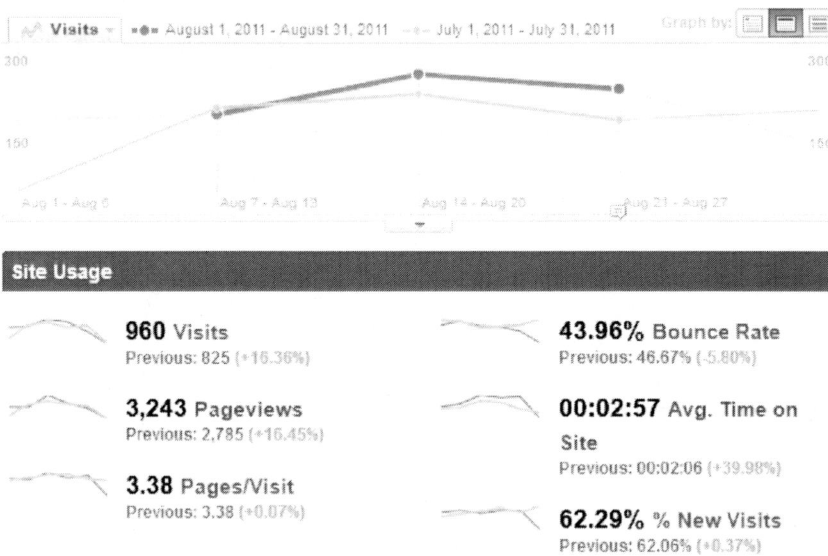

FIGURE 3: Effect of product descriptions, content and metadata improvement on website visitations.

A relative simple update to the content, and ensuring that the metadata was reflective of the new content, produced a significant increase in website traffic. This was mostly because of the presence of relevant, unique, keyword rich and regularly updated content. This increase in traffic confirms this conclusion and website authors should seriously consider this crucial aspect of a website.

OVERALL TRAFFIC TRENDS

Month 1

Next, two one-month periods are compared to determine whether a trend is evident. See Figure 4.

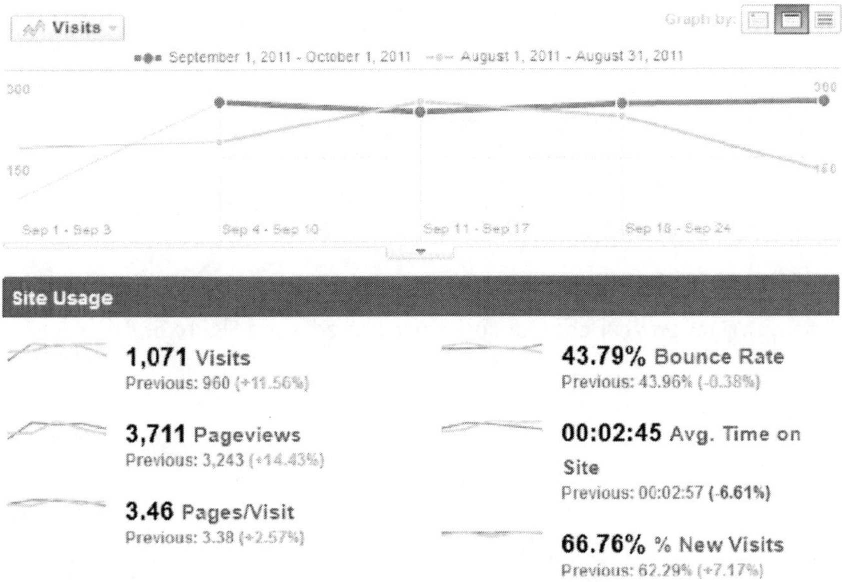

FIGURE 4: Total website traffic: September 2011 compared to August 2011.

During the month of September 2011 no work was conducted on the test website. In this time, the total traffic increased further compared to August 2011; 11.56% in total. Although August 2011 had not quite reached the traffic levels of May 2011, which included the PPC traffic, the data looked promising. Also note that August 2011 had two calendar events that would have negatively impacted the results for August 2011, as noted earlier. Furthermore, September 2011 also had one event on 24 September 2011 when South Africans celebrated Heritage Day.

By this time it was safe to assume that Google had indexed all the on-site changes. It was evident that all the website's pages were being re-indexed and ranked. However, the continuing climb in traffic indicated that rankings were indeed improving and more and more users were finding the website via Google's search result pages. This has to be seen against the fact that no further investment in PPC was done after campaign shutdown in May 2011.

Month 2

In October 2011 the authors continued to monitor the results and noted another slight increase in total traffic (1.21%). This was an indication that Google had completed the re-indexing of the modified test website. By this time the website traffic had increased from 825 visits to 1084 (31.4%

increase) in a period of four months. It should be noted that, in October 2011, the schools in South Africa closed for recess from 01 October 2011 to 10 October 2011. This would have had a negative impact on the October 2011 data. Also, as mentioned before, Heritage Day was celebrated on 24 September 2011.

Annual comparison

Finally, a year on year comparison is performed – see Figure 6.

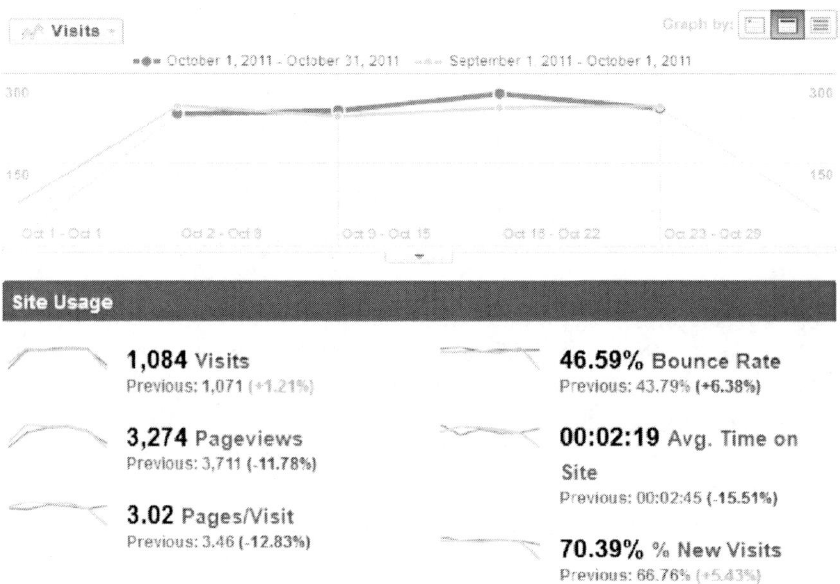

FIGURE 5: Total Traffic October 2011 compared to September 2011.

In Figure 6 the authors compare the period 01 June 2011 to 31 October 2011 (no PPC included) to 01 June 2010 to 31 October 2010 (PPC traffic included). In June, the data regarding website traffic for the two years appear similar, however a sharp decrease in traffic can be observed when the PPC campaign was shut down and the modifications to the website started. By August 2011 the traffic already started increasing to just below the levels of the previous year after the header tags, image file names and image ALT tags were modified. Then, in August 2011 when the product descriptions, new homepage content, and metadata were concluded the traffic increased further. By end of September 2011 the total traffic had increased to levels higher than the previous year, which included the PPC traffic.

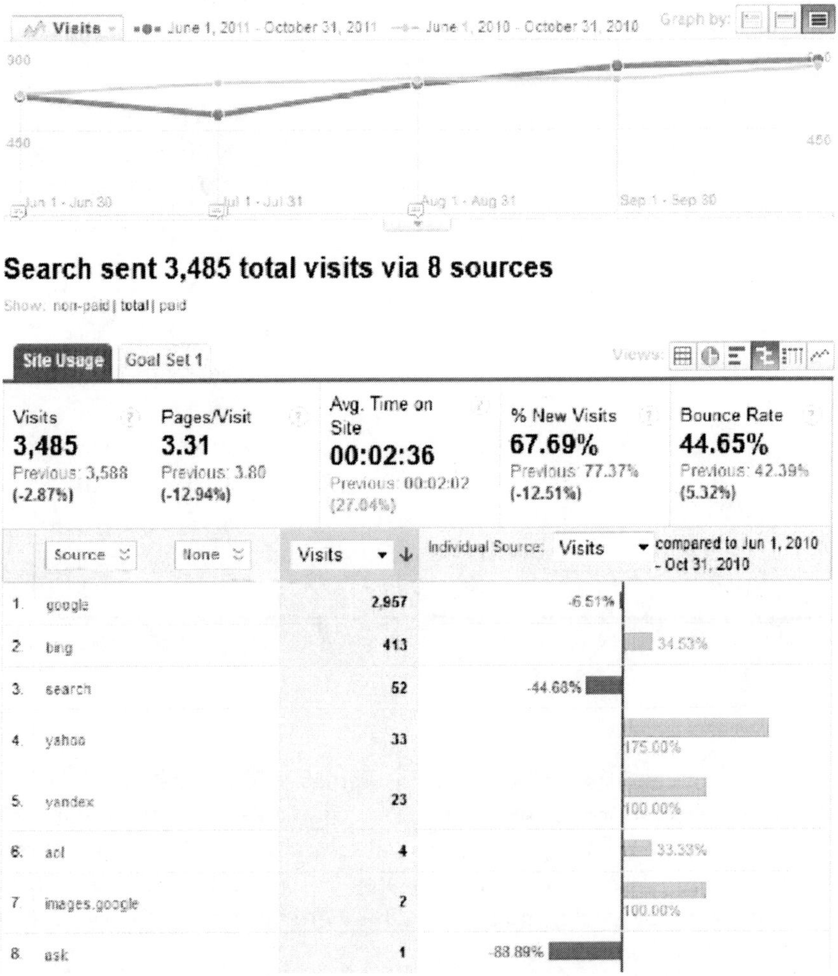

FIGURE 6: Year on year comparison: 01 June 2011 – 31 October 2011 compared to 01 June 2010 – 31 October 2010.

Final traffic or expenditure summary

The results of this study are best described by summarising traffic to the XYZ website during the two test periods, whilst comparing it to the expenditure. Table 1 summarises the relevant figures.

TABLE 1: Comparative traffic during two periods.

Date	Traffic PPC	Traffic SEO	Exp. Adj.	Exp. Orig.
Jun-10	650	-	142.8571	3000
Jul-10	706	-	142.8571	3000
Aug-10	723	-	142.8571	3000
Sep-10	721	-	142.8571	3000
Oct-10	788	-	142.8571	3000
Nov-10	718	-	142.8571	3000
Dec-10	718	-	142.8571	3000
Jan-11	718	-	142.8571	3000
Feb-11	718	-	142.8571	3000
Mar-11	718	-	142.8571	3000
Apr-11	718	-	142.8571	3000
May-11	718	-	0	0
Jun-11	-	639	904.7619	19000
Jul-11	-	536	0	0
Aug-11	-	696	0	0
Sep-11	-	792	0	0
Oct-11	-	822	0	0

PPC, pay-per-click; SEO, search engine optimisation; Exp. Adj.; Expenses Adjusted; Exp. Orig.; Expenses Original'.

The date column lists the date of measurement of both traffic values and the expenditure of company XYZ on SEM. 'Traffic PPC' lists the number of visitors for the period originating from all search traffic; direct, referral and PPC sources. However, the actual figures for November 2010 until May 2011 were not available, and these figures were taken to be the average of the preceding months (718).

The 'Traffic SEO' column contains the number of visitors for the relevant period, all of which is search traffic. However, by this time the PPC campaign had been terminated, so part of the source is SEO-generated traffic. The 'Expenses Original' column is a summary of the SEM expenditure of the XYZ Company, being R3000 per month spent on PPC for 11 consecutive months. In May 2011 there was no expenditure, since the PPC campaign was terminated. Then, in June 2011, an amount of R19 000 was paid (once-off) to do the SEO as agreed.

Finally, the 'Expenses Adjusted' column is simply the 'Expenses Original' figure divided by 21. The figure of 21 was chosen as a scale-down factor to bring the Rand values in line with the other figures in the table. This would ensure that graphs of these values plotted on the same scale (see Figure 7) would be comparable in amplitude.

A graphical presentation of the relevant figures as noted above is done in Figure 7.

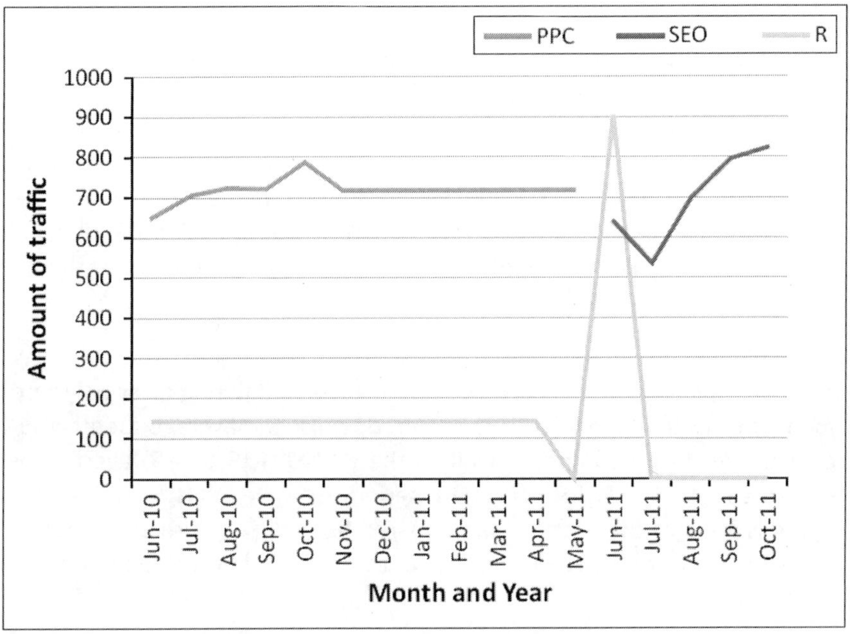

FIGURE 7: Summary of traffic versus expenditure.

An analysis of the graphs in Figure 7 is required. The PPC graph indicates that the amount of search traffic had stabilised over the 12 month period from June 2010 to May 2011. Since this component consists of three parts and the split is unknown, it cannot be determined what the exact contribution of PPC was to the monthly figure. However, it is assumed that it must have been substantial considering the steep drop in traffic in the months (June and July 2011) directly following the termination of the PPC campaign.

The SEO graph of Figure 7 again lists all search traffic to the website, but this time with the PPC component having been replaced by the newly introduced SEO component. The initial drop over the first two months is owing to the SEO campaign taking time to influence rankings and therefore traffic. However, as from the third month into the SEO section, the graph shows a continued rise indicating a growing trend in traffic volumes. This resulted from the fact that search engine crawlers had started visiting the site, indexing the new content, and their algorithms had started giving the site an ever increasing rank in the result listings. The rising trend appears to flatten towards the last recorded month, but without more figures this trend cannot be confirmed past the last recorded month.

Finally, the expenses of XYZ to achieve this increase in traffic need to be put into perspective – refer to the 'R' line on Figure 7. At a fixed rate of R3000 per month, XYZ had spent R33 000 over the 11 month period in an attempt to maximise the number of visitors to their site. This expenditure played a large part in achieving the traffic volumes listed as the PPC line as noted above. During June 2011 there was a sharp increase to R19 000, but this was a once-off expenditure. After that time it dropped to zero for the rest of the recording period.

It should be noted that the traffic volume during the SEO period had bypassed the highest level achieved during the PPC period after only three months of running on SEO. This occurred whilst the average monthly PPC expense of R3000 compared well to the R3800 (R19 000/5) average per month spent on SEO for this short period. However, this R3800 average figure is for only five months as compared to the PPC figure of 11 months. The average cost for SEO, if calculated further, would decrease dramatically as follows:

- November 2011: R3167 per month
- December 2011: R2714 per month
- May 2012: R1727 per month

In both cases, after 11 months the PPC expense would still be R3000 per month with the SEO averaging out at R1727 per month. This trend of downward spiralling costs for SEO would continue. At the same time, if the SEO graph is extrapolated, it can be expected that the traffic volumes could increase past those achieved during the PPC period.

It should be noted that in a real-life SEO campaign there will be further expenses past the initial layout, but not necessarily of the monthly recurring type. In the time period of this case study there were none, bar the initial amount. However, a study over a much longer period of time is needed to do a long-term comparison. It is likely that the longer an SEO campaign is running, the lower the average monthly costs will become.

Limitations of this study include that the data were obtained from a relatively low-traffic site. Also, data across a longer period of time would provide a clearer trend. Finally, a comparison was done on the traffic generated from certain marketing expenses only, not the profit eventually generated by that traffic.

CONCLUSION

An existing commercial website (that of company XYZ) was considered as the central object of this research project. An experiment was done on the XYZ website to start implementing SEO immediately after the PPC was terminated. At the same time, monitoring was done of traffic to the website and of other relevant analytic measures. The SEO was done based on industry best practice supported by academic research.

The owners had been spending 'marketing dollars' on this website, using firstly, only one of the two main types of SEM for a period: PPC. The main disadvantage of PPC is that expenditure has to be consistent for results to appear; the moment a PPC budget is cut, the resultant traffic drops to zero immediately. Therefore, the overall PPC expenditure should be calculated from the monthly expenditure; in the case of XYZ, R3000.

After the PPC campaign was terminated, XYZ spent R19 000 once-off on SEO. This amount was for the implementation of the elements as listed under 'Methodology' above. As noted from the literature, where PPC produces immediate ranking improvements once payment has been arranged, SEO takes longer to affect rankings. This is mostly due to the waiting period for search engine crawlers to visit or refresh their copy of the website content.

When these expenditure figures are compared, it can thus be claimed that, after 6.33 months, the expenditure on the two systems for XYZ would have been the same. From this point in time onwards, SEO would continue to provide a growing return on investment over PPC, assuming that no further expenses would be required for SEO.

In conclusion, it can be claimed that, in this specific case, SEO provides a better investment than PPC. It is also predictable that this advantage will increase as time goes by. It should also be noted that traffic to a website on its own is not the only indicator of success. A high conversion rate, leading to more revenue generated, and eventually leading to increased profit would be the final indicator of the success of an e-commerce website. However, owing to the sensitivity of company financial information, this kind of evidence will be harder to come by and use as proof.

COMPETING INTERESTS

The authors declare that they have no financial or personal relationships which may have inappropriately influenced them in writing this article.

AUTHORS' CONTRIBUTIONS

Both W.T.K. (Cape Peninsula University of Technology) and M.W. (Cape Peninsula University of Technology) worked together in conceptualising the research work, completing the literature survey and reference hunting, doing the write-up and checking results. W.T.K. gathered and summarised the data, checked accuracy and did the write-up of the skeleton of the whole article, specifically the results section. M.W. did the conclusion, abstract, proofreading, editing and overall checking.

REFERENCES

1. Akakandelwa, A., 2011, 'An exploratory survey on the SADC e-government web sites', *Library Review*60(5), 421–431.

2. Barreto, A.M., 2013, 'Do users look at banner ads on Facebook?', *Journal of Research in Interactive Marketing* 7(2), 119–139.

3. Boyes, J.A. & Irani, Z., 2004, 'An analysis of the barriers and problems to web infrastructure development experienced by small businesses', *International Journal of Information Technology and Management*3(2/4), 189–207.

4. Brunn, D.S. & Dodge, M., 2001, 'Mapping the "worlds" of the World Wide Web: (Re) structuring global commerce through hyperlinks', *American Behavior Scientists* 44(10), 1717–1739.

5. Chen, C.-Y., Shih, B.-Y., Chen, Z.-S. & Chen, T.-H., 2011, 'The exploration of internet marketing strategy by search engine optimization: A critical review and comparison', *African Journal of Business Management*5(12), 4644–4649.

6. Craven, T.C., 2003, 'HTML tags as extraction cues for webpage description construction', *Information Science Journal* 6, 1–12.

7. Crowley, M., 2014, 'Take the guesswork out of your marketing strategy', *Journal of Financial Planning*27(1), 16.

8. Green, D., 2000, 'The evolution of Web searching', *Online Information Review* 24(2), 124–137.

9. Guenther, K., 2014, 'Getting your website recognized', *Online Magazine*, Issue May/June, 47–49. *Guidelines for optimising and making your site searchable*, 2014, viewed 01 March 2015, fromhttp://www.ananzi.co.za/submit/guidelines

10. Gupta, A. & Mateen, A., 2014, 'Exploring the factors affecting sponsored search ad performance', *Marketing Intelligence* & *Planning* 32(5), 586–599.

11. Henzinger, M.R., Motwani, R. & Silverstein, C., 2002, 'Challenges in web search engines', viewed 30 April 2015, from http://www.acm.org/sigs/sigir/forum/F2002/henzinger.pdf

12. Hubbard, J., 2004, 'Indexing the Internet', in *Essay for Drexel University College of Information Science and Technology*, viewed 25 April 2015, from http://www.tk421.net/essays/babel.html

13. Kennedy, K. & Kennedy, B.B., 2008, 'A small company's dilemma: Using search engines effectively for corporate sales', *Management Research News* 31(10), 737–745.

14. Kent, P., 2012, *Search engine optimization for dummies*, 5th edn., Wiley and Sons, Hoboken.

15. Kisiel, R., 2010, 'Dealers get on top of search engine results', *Automotive News* 84(6408), 24–25.

16. Kumar, E. & Kohli, S., 2007, 'A strategic analysis of search engine advertising in web based-commerce',*Journal of Internet Banking* & *Commerce* 12(2), 1–13.

17. Lee, W.I., 2010, 'The development of a qualitative dynamic attribute value model for healthcare institutes',*Iranian Journal of Public Health* 39(4), 15–25. PMID: 23113034.

18. Lee, W.I., Chen, C.W. & Wu, C.H., 2010, 'Relationship between quality of medical treatment and customer satisfaction - A case study in dental clinic association', *International Journal of Innovative Computing, Information and Control* 6, 1805–1822.

19. Lee, W.I. & Lin, C.H., 2011, 'Consumer hierarchical value map modeling in the healthcare service industry',*African Journal of Business Management* 5(3), 722–736.

20. Malaga, R.A., 2010, 'Search engine optimization - Black and white hat approaches', *Advances in Computers*78, 1–39.

21. Mancuso, A., 2013, 'How Google AdWords' Ad Rank Algorithm Update Increased Brand CPC's by 600 Percent', viewed 12 May 2015, from http://www.seerinteractive.com/blog/how-google-adwords-ad-rank-algorithm-update-increased-brand-cpcs-by-600

22. Moreno, L. & Martinez, P., 2013, 'Overlapping factors in search engine optimization and web accessibility',*Online Information Review* 37(4), 564–580.

23. Moxley, D., Blake, J. & Maze, S., 2004, 'Web search engine advertising practices and their effect on library service', *The Bottom Line: Managing Library Finances* 17(2), 61–65.

24. Murphy, H.C. & Kielgast, C.D., 2008, 'Do small and medium-sized hotels exploit search engine marketing?',*International Journal of Contemporary Hospitality Management* 20(1), 90–97.

25. Nabout, N.A., 2015, 'A novel approach for bidding on keywords in newly set-up search advertising campaigns', *European Journal of Marketing* 49(5/6), 668–691.

26. Nielsen, J., 2014, 'Top ten global web parent companies', viewed 08 March 2015, from

27. Olbrich, R. & Schultz, C.D., 2014, 'Multichannel advertising: Does print advertising affect search engine advertising?', *European Journal of Marketing* 48(9/10), 1731–1756.

28. Onaifo, D. & Rasmussen, D., 2013, 'Increasing libraries' content findability on the web with search engine optimization', *Library Hi-Tech* 31(1), 87–108.

29. Oppenheim, C., Morris, A. & McKnight, C., Lowley, S., 2000, 'The evaluation of WWW search engines',*Journal of Documentation* 56(2), 190–211.

30. Plaza, B., 2009, 'Monitoring web traffic source effectiveness with Google Analytics: An experiment with time series', *Aslib Proceedings* 61(5), 474–482.

31. Sagin, E., 2013, 'What the new AdWords ad rank algorithm really means', viewed 29 January 2015, fromhttp://www.wordstream.com/blog/ws/2013/10/24/adwords-ad-rank-algorithm

32. Sen, R., 2005, 'Optimal search engine marketing strategy', *International Journal of Electronic Commerce*10(1), 9–25.

33. Singh, A.M., 2002, 'The Internet – Strategies for optimal utilization in South Africa', *South African Journal of Information Management* 4(1).

34. Thelwall, M., 2001, 'Commercial web site links', *Internet Research* 11(2), 114–124.

35. Vaughan, L., 2014, 'Discovering business information from search engine query data', *Online Information Review* 38(4), 562–574.

36. Visser, E.B. & Weideman, M., 2011, 'An empirical study on website usability elements and how they affect search engine optimisation', *South African Journal of Information Management* 13(1). 9 pages.

37. Weideman, M., 2013, 'Comparative analysis of homepage website visibility and academic rankings for UK universities', *Information Research* 18(4), viewed 30 April 2015, from http://InformationR.net/ir/18-4/paper599.html

38. Weideman, M., 2009, *Website visibility: The theory and practice of improving ranking,* Chandos Publishers, Oxford.

39. Zuze, H. & Weideman, M., 2013, 'Keyword stuffing and the big three search engines', *Online Information Review* 37(2), 268–286.

Chapter 10

Information and Communications Technology Adoption Amongst Township Micro and Small Business: The Case of Soweto

Carl Marnewick[1]

[1]Department of Applied Information Systems, University of Johannesburg, South Africa

ABSTRACT

Background: No empirical evidence is available on whether or not information and communications technology (ICT) is adopted by South African township micro and small businesses (MSBs) in order to grow the business. ICT, as an enabler, can grow the business. It is evident from literature that ICT enables MSBs to be more productive and efficient.

Objectives: This research determines to what extent ICT is adopted by MSBs in Soweto, and whether or not a positive correlation exists between the adoption rate of ICT and the financial and growth performance of the MSB itself.

Method: A structured questionnaire was completed by 978 respondents to determine the extent of ICT adoption and for what ICT is used. This exploratory research provides new knowledge about the acceptance of ICT within township MSBs.

Results: The results indicate that ICT is not used to the fullest by MSBs. Rather, it is used as a basic tool for doing business but it does not form an integral part of the business. This research provides insight into the usage and adoption of ICT and it opens the door for further cross-analysis research.

Conclusion: Education and training are needed to ensure that MSBs use ICT to the fullest. MSBs embracing ICT can evolve from a survivalist SME to a more sustainable micro and small MSB.

INTRODUCTION

Small and medium enterprises (SMEs) are the heartbeat of any country's economy. This is even more so in developing countries where SMEs are responsible for employment and the growth of the economy (Beck, Demirguc-Kunt & Levine 2005; Cravo, Gourlay & Becker 2012). One aspect that can assist SMEs in growing the business is information and communication technology (ICT). Management literature highlights the value of ICT and information in the organisational context (Hu & Plant 2001). ICT can enable SMEs to achieve these goals, of employment and growth, if it is adopted and incorporated into the day-to-day management of the SME.

There is currently no empirical evidence of whether or not micro and small businesses (MSBs), as a special type of SME, in townships adopt any type of ICT to grow the business. The term township originated in the apartheid era and began when non-white people (black people, people of mixed-race and Indians) were limited to living near or working in white-only communities. Township MSBs play a major role in the creation of wealth and employment within the realm of townships, as there are no other forms of employment in those areas (Mbonyane 2006). According to SBP (2013):

The contribution that small business makes in generating employment is near universally recognised. So is the corollary acknowledgement, that a healthy and much expanded small business community could be a keystone of a new economic deal for South Africa. (n.p.)

The argument is that ICT, as a business enabler, should be capable of assisting MSBs if it is adopted by the MSBs. ICT provides an opportunity for MSBs to improve their efficiency and effectiveness, and even to gain competitive advantage (Arendt 2008; Thong & Yap 1995). Research carried out, on rural SMEs in the United States of America (USA) in the late 1990s, concluded that SMEs will only adopt ICT if there is a perception that ICT will assist in overcoming a business problem or opportunity (Premkumar &

Roberts 1999:470). Some of the benefits mentioned by the SMEs that had adopted ICT were reduced turnaround time, increased transaction speed, access to current information and reduction in data entry errors (Premkumar & Roberts 1999:482). There are, however, SMEs that deliberately did not adopt any ICT. One of the reasons was that the SMEs did not find any need for these technologies in their businesses.

This research is innovative as it highlights that township MSBs do not adopt ICT and that interventions are needed to cross the digital divide in townships. The results provide statistical information on the 'raw' rates of ICT adoption for a sample of MSBs in the township of Soweto. This research also provides a baseline for future research. Future research can include comparative analysis of various MSBs as well as SMEs in various townships, and analysis of developed and developing countries.

Studies within the discipline of ICT adoption have not focused on SMEs or, for that matter, MSBs within townships. Instead, research has been focused on traditional companies and SMEs and how ICT enabled those organisations to perform financially (Lucchetti & Sterlacchini 2004; Premkumar & Roberts 1999; Thong & Yap 1995).

This research addresses the lack of empirical knowledge with regard to ICT adoption and usage within township MSBs. An understanding of the problem provides various possibilities for South Africa. Firstly, the adoption of ICT will increase the revenue of the individual MSBs and ultimately the growth of the economy at large. Secondly, it also provides focus areas where ICT and the type of ICT can and should be deployed to assist MSBs.

A structured questionnaire was circulated amongst MSB owners within Soweto. This was circulated through a survey consultancy company that interviewed the population. This company deployed resources, the personnel of which interviewed the various SMB owners. The interviewers were briefed on the purpose of the research, and additionally about how each question related to the overall research hypotheses. The data were then captured and analysed using SPSS.

The article is divided into four sections. The first section focuses on the literature of MSBs, as well as the technology adoption model (TAM), and how it can and should assist MSBs in the growth of the economy. The second section focuses on the research methodology and how the results were collected from the various respondents. The third section is an analysis of the results of the 978 respondents, and focuses on the type of ICT that the MSBs used as well as the business reasons that the ICT was used. The fourth and last section specifies the impact of ICT adoption and the consequences of adopting or not adopting ICT within the realm of township MSBs.

Literature review

The value of the small business sector is recognised in economies worldwide which is irrespective of whether it is a developed or a developing economy (Thurik & Wennekers 2004; Hotho & Champion 2011). According to Mahembe (2011:13): 'the contribution towards growth, job creation and social progress is valued highly and small businesses are regarded as an essential element in a successful formula for achieving economic growth'. The contribution of SMEs in South Africa to the economy is summarised as per Table 1. The World Bank (2006), however, estimated that SMEs would contribute 39% to the generation of employment whereas China's SME sector contributed 78% to its total employment. This is an area of concern given the high unemployment rate in South Africa. SMEs can bridge this gap.

TABLE 1: Contribution of South Africas small and medium enterprises to the economy.

Enterprises	Number of firms (%)	Employment (%)	GDP (%)
Survivalist	19.6	2.2	
Micro (0)	31.3	3.5	5.8
Micro (1–4)	19.8	6.5	
Very small	20.5	13	13.9
Small enterprises	6.8	15.7	
Medium enterprises	1.3	13	15
Large	0.7	46.1	65.2

According to Mandisa Mpahlwa, the former South African Minister of Trade and Industry, the small business sector in 2006 comprised 2 million small businesses which present almost 98% of the total number of South African companies (Njiro, Mazwai & Urban 2010). Collectively SMEs employed 55% of the country's labour force. However, 87% of these small businesses are surviving on a day-to-day basis and they are owned by black people. Women owned 41% of these. Trends such as these are very significant, as they illustrate both the potential and the challenges for small business growth (Njiro*et al.* 2010).

SMEs are normally defined by three components i.e. by revenue, assets (excluding fixed property) or the number of employees. Each country has varying cut-off points and these may even vary by industry. South Africa's National *Small Business Act* (Act No. 102 of 1996) defines SMEs as a 'separate and distinct business entity', and this includes cooperatives and non-governmental organisations (Kennedy, Bounds & Goldman 2011).

This *Act* uses the number of employees, the size as well as the annual turnover as measurement criteria. SMEs can be classified as follows:

- Survivalist enterprise: The income that is generated is less than the minimum income standard. Survivalist enterprises include hawkers, vendors and subsistence farmers. Survivalist enterprises are often categorised as part of the micro enterprise sector.

- Micro enterprise: The turnover is less than the stipulated South African value-added tax (VAT) registration limit. These enterprises do not formally register the enterprise and include, for example, *spaza* shops, minibus taxis and household industries. Micro enterprises employ no more than five people. The word '*spaza*' means 'camouflaged' or 'hidden' in township slang and, therefore, '*spaza* shop' refers to a camouflaged or hidden shop in a township (Terblanche 1991:37).

- Very small enterprise: They employ fewer than 10 employees, except for the mining, electricity, manufacturing and construction sectors, in which they employ fewer than 20 employees. These enterprises operate in the formal market and have access to ICT.

- Small enterprise: They employ up to 50 employees. Small enterprises exhibit more complex business practices.

- Medium enterprise: They employee a maximum number of 100 employees, or up to 200 employees for the mining, electricity, manufacturing and construction sectors. These enterprises might have additional management layers (Abor & Quartey 2010:221).

Another way to classify SMEs is based on the classification methods of the Bolton Committee, which classifies and describes SMEs on the basis of sectors (Abor & Quartey 2010). In the manufacturing, mining and construction sectors, it includes all companies that have 200 or fewer employees. In the road transport industry, companies are considered to be small when the number of vehicles is five or fewer. Companies operating in retail, wholesale or services are measured on the basis of their annual turnover.

The South African classification of SMEs is substantially different from that of the European definition. SMEs, as per the European definition, imply companies with less than 250 employees (Silvius 2004). The factors that mainly determine if a company is an SME are displayed in Table 2 (European Commission 2012).

TABLE 2: Factors determining small and medium enterprises status.

Company category	Employees	Turnover
Medium-sized	< 250	≤ € 50 million
Small	< 50	≤ € 10 million
Micro	< 10	≤ € 2 million

An SME in South Africa's SME sector is required to fulfil the following:

[*They are*] expected to fulfil a number of roles, ranging from poverty alleviation and employment creation to international competitiveness. Not only are these very divergent policy objectives, but the policy instruments introduced to meet these objectives can be equally different, ranging from literacy training to technological advice. (Blueprint Strategy & Policy (Pty) Ltd 2005:9)

It is now widely accepted that small businesses are the chief contributor to job creation worldwide and this trend is also true of South Africa, as per Table 1.

In South Africa, SMEs contribution is 56% to the employment in the private sector and 36% to the gross domestic product (Olawale & Garwe 2010:729). They also state that small businesses are one of the best ways to address unemployment through the potential of the small business itself to create employment and to promote the development of small businesses.

SMEs also play an important role within townships to uplift the local community and to create jobs. Townships are a particularly South African phenomena and a deliberate manipulation of former urban planning that designed cities in terms of race. The creation of Soweto, which is a conglomeration of townships, was founded to house mainly black labourers who worked in mines and other industries in the city of Johannesburg. An increasing number of black people were relocated from inner city 'black spots' to Soweto, as the centre of Johannesburg was reserved for white people (Njiro *et al.* 2010:6). In 1963, after a public four-year competition, the name Soweto was adopted for the South Western Townships (Njiro *et al.*2010:6). Apartheid planning did not provide any infrastructure to these settlements and all manner of shelters, including shacks, flourished. Residents had limited access to capital, education and basic social and economic rights. They were also prohibited from owning and running businesses.

According to Njiro *et al.* (2010):

Of the 200 000 Soweto businesses, about 83% have turnovers of less than R10 000 a month. The objective is to grow these small businesses into entities with turnovers exceeding R1 million. (p.8)

To achieve this goal of growing MSBs from survivalist and micro enterprises into very small and small enterprises, ICT can be used. ICT is seen as an enabler for business (Mitra 2005). The usage of ICT can reduce the cost of operations and increase profits. This is applicable to any enterprise, irrespective of size. ICT can help MSBs to cut costs through the improvement of internal business processes, improved communication with customers as well as suppliers and an improved supply chain online (Kannabiran & Dharmalingam 2012:192). ICT can also help MSBs to manage, source, produce, plan and ultimately support decisions in their daily operations.

This is not always the case, however. ICT deployment in many Chinese MSBs resulted in negative performance which is not in parallel with the economic growth that was experienced during the same period. The critical challenges were not caused by the failure of ICT technologies. Rather, inadequate support and business processes were common obstacles in the deployment of ICT (Sidney, Tina & Choy 2010). Another case in point is Indonesia, where MSBs are one of the economic backbones. In their daily operation some of the MSBs conduct their business using ICT. However, the majority do not use ICT in an optimum way and daily operations are still performed manually (Santosa & Kusumawardani 2010).

MSBs adopt ICT to achieve efficiency and in response to a variety of environmental and internal pressures (Khalifa & Davison 2006:275). Moreover, the MSBs most likely to use ICT are very likely to be technically competent already. MSBs do not adopt ICT uniquely as a response to coercive and normative pressures from external and internal sources. They are also strongly motivated by the perceived desirability and feasibility of ICT under consideration.

The literature review indicates that there is a case to be made for the adoption and use of ICT within the day-to-day running of an SME. The second part of the literature review focuses on the current models which explain when and why MSBs would adopt and use ICT.

Davis (1989) emphasises that the (1) perceived usefulness and the (2) perceived ease of use determines whether someone will adopt technology or not. This thinking realised the benefits of the technology acceptance model (TAM). This model was further refined by Davis, Bagozzi and Warshaw (1989), who included (1) the attitude towards using and (2) the behavioural intention to use additional aspects of TAM. Figure 1 illustrates TAM.

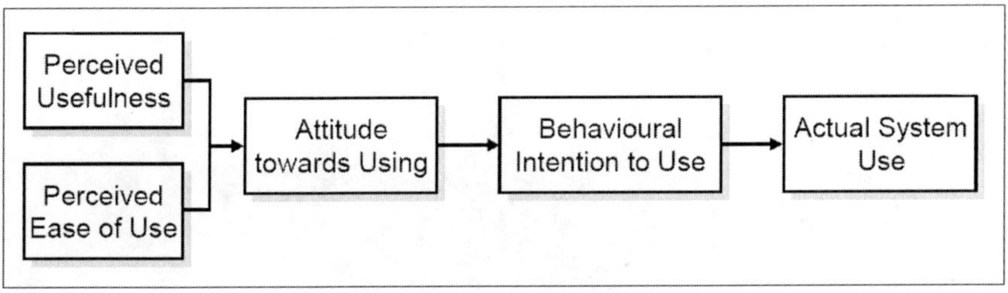

FIGURE 1: Technology acceptance model.

Various other models have emerged since then, but all of them basically incorporate the original components of TAM (Venkatesh, Morris, Davis & Davis 2003). To relate this back to MSBs in Soweto, MSBs will only adopt ICT if there is a perception that ICT will assist in the growth and sustainability of the MSB itself and if the perception exists that it is easy to use a specified technology.

Lucchetti and Sterlacchini (2004) classify the adoption of ICT in SMEs into three categories:

- General-use information and communication technologies (ICTs) include collaboration tools where the rate of adoption is high and do not depend on size (i.e. number of employees) and industry.
- Production-integrating ICTs include local area network (LAN), electronic data interchange (EDI) and intranet. This type of ICTs is linked to intra

and inter-firm production processes; they are more expensive than general-use ICTs and are dependent on relevant technological skills.

- Market-oriented ICTs are jointly identified by the presence and the content of a firm's web site and are used to improve visibility and provide detailed information of products, with a view to attract potential customers (p. 152).

This is all well and good in the context of a developed nation, but access to ICT in Soweto is not a given. The results from the South African 2012 census highlight the digital divide. The 'proportion of households owning cellphones increased from 31.9% in 2001 to 88.9% in 2011' (Statistics South Africa 2012). The proportion of households with computers increased from 8.5% to 21.1% between 2001 and 2011, respectively (Statistics South Africa 2012). Figure 2 shows the percentage of South African households that have access to the Internet. A relatively high proportion of these households, 64.8%, have no access to the Internet.

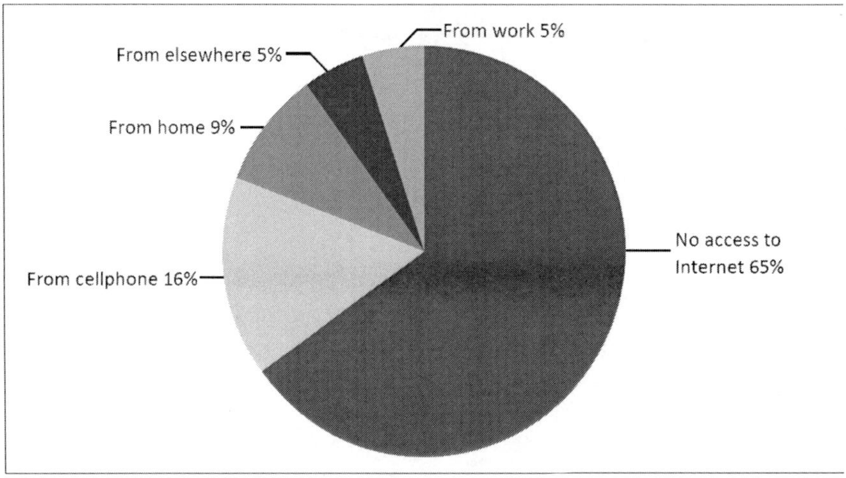

FIGURE 2: Internet access.

The highest percentage of households reported that cellphones are used to access the Internet (Statistics South Africa 2012). A report on the status of ICT in Africa highlighted the challenges that MSBs face. The main challenges are infrastructure and a growing ICT skills gap (Yonazi *et al.* 2012).

The review of the literature highlighted two important aspects: (1) MSBs as well as SMEs are important for any country's economy and (2) the adoption of ICT itself can improve the profitability of an MSB.

This leads to the research problem at hand: is ICT, per se, adopted by township MSBs in order to improve the profitability and sustainability of the MSB?

The research is of an exploratory nature and, therefore, the focus is on the actual use of ICT and not necessarily on the causes of technology acceptance. Given the model in Figure 1, the focus is not on the factors that influence the use of ICT but on the actual adoption of ICT itself. The reason for this is that one first needs to understand a concept before it can be explained.

Given this rationale, the following research questions are posed:

1. What type of ICT is adopted by MSBs in Soweto?
2. Is the adoption of ICT dictated by the type of MSB?
3. Is there a correlation between the type of ICT adopted and the profitability of the MSB?
4. Is ICT applied differently based on the type of MSB?

The following section contains a discussion of the research methodology that was followed to gather the responses.

Research methodology

Research designs focus on validating the research question through a testing project and can be either quantitative research designs or qualitative research designs (Feilzer 2010; Huizingh 2007). Quantitative research focuses on answering the research question through the collection of numerical data and the statistical analysis of this data (Balnaves & Caputi 2001; Blaikie 2003).

A structured questionnaire was developed focusing on two aspects: (1) biographical information such as income and type of SME and (2) the type of ICT used, for instance cellphones, as well as the frequency of usage. The types of ICT selected are founded on research, indicating that mobile technology is a main enabling technology (Ngassam, Ntawanga & Eloff

2013). Building on mobile technology, the next levels of technology are personal computers (PCs), laptops and tablets.

Structured questionnaires make use of closed questions where the results can be analysed quantitatively for patterns and trends. The researcher opted for a structured questionnaire because it ensures that each respondent is presented with exactly the same questions in the same order. This ensures integrity in longitudinal studies as well as comparisons between various subgroups.

TAM, which informs this research, uses 28 questions to measure the perceived usefulness (14 questions) and perceived ease of use (14 questions). The intention is to adopt the TAM questions in a specific environment and for a specific application (Polun *et al.* 2004). The results can then be used to measure and correlate the perceived usefulness and ease of use. The problem that was faced in Soweto is that there are no data available on the acceptance of ICT and, therefore, the ease of use and usefulness of a specific technology is of no relevance at this point in time. It is important to understand, within the context of Soweto and its ICT adversity, what technologies are currently used as well as the purposes for which they are used. The 28 questions of TAM could, therefore, not be used as they are normally worded, and a shortened structure questionnaire, derived from TAM, was designed. The questions were grouped into two major sections focusing, firstly, on the purpose of the respective technologies and, secondly, on how frequent these technologies are used. The rationale is that the results will then be used to develop in-depth TAM questionnaires for each of the technologies, once there is an understanding of what technologies are used and how frequently.

The questionnaires were administered via a private consulting firm that was familiar with Soweto and the MSBs within the different suburbs. The consulting firm made use of runners to interview the SME owners. The runners were used as a pilot to test the questionnaire to identify potential problems and to make improvements. Using the runners as a pilot scheme, also made them aware of the type of responses that they could expect.

A briefing session was held to explain the questions to the runners, and the types of answers that should be provided by the SME owners. Soweto was divided into its suburbs (33 in total) and 30 SME owners were picked randomly to complete the questionnaire. The owners were chosen on an availability basis as the runners surveyed the suburb. The language of choice used was either English or the native language of the SME owner which consisted, in most cases, of Zulu. The contact details of the SME were also added for follow-up questionnaires and future comparative analysis. A total of 978 completed questionnaires were received. The quality of the responses was of such a nature that all the questionnaires could be used.

The data gathered in this survey were processed and analysed using SPSS, a statistical analysis software package (Argyrous 2011; Huizingh 2007). SPSS is able to perform enhanced data management and has extended reporting capabilities.

Reliability is concerned with how well the survey data can be reproduced, and validity is a measure of how well the survey measures what it is supposed to measure (Kirk & Miller 1986; Kitchenham & Pfleeger 2002; Litwin 1995). Some available reliability checks that can be done in a survey are presented by Litwin (1995):

- Test-retest is relevant when a survey is sampled at different points in time (Kitchenham & Pfleeger 2002). In this research the survey was sampled for the first time; therefore, a test-retest is only possible in future.

- As scales were used in the questionnaire, it was very important to check internal consistency. Internal consistency is a measure when assessing scales are used in the survey (Litwin 1995). Cronbach's alpha is an internal consistency measure and is used to determine whether the scale is reliable or not (Kitchenham & Pfleeger 2002; Litwin 1995). The Cronbach's alpha coefficient is a reflection of how well the different items complement one another in measuring the same variable (Cronbach 1951; Litwin 1995). The Cronbach's alpha coefficient was calculated for each of the technologies that were used and is presented in Table 3. Internal consistency levels of 0.7 or more are generally accepted as representing good reliability.

The results indicate that there is internal validity apart from the cellphone grouping of questions. The results of this grouping must be interpreted carefully and a retest might highlight some additional inconsistencies.

The purpose of any questionnaire is that it measures what it intended to measure (Cameron & Price 2009). If a questionnaire does not measure what it is supposed to measure, then the conclusions and statistical analysis might also be invalid. Validity checks are available to verify that the questionnaire is suitable (Kitchenham & Pfleeger 2002). The types of validity that can be used to assess the survey questionnaire are face, content, criterion and construct validity (Litwin 1995).

TABLE 3: Cronbachs alpha reliability tests.

ICT adoption	Cronbach's alpha	Cronbach's alpha based on standardised items	*N* of items
Cellphone	.570	.585	6
Smartphone	.881	.876	7
PC	.796	.837	6
Laptop	.792	.845	6
Internet	.820	.836	4

The researcher opted for face and content validity. Face validity refers to the 'obviousness' of a test that is the degree to which the purpose of the test is apparent to those taking it. The purpose of this research, as well as the questions, was clear to the respondents and, thus, it can be derived that there is high face validity. Content validity on the other hand focuses on extent to which the items are fairly representative of the entire domain the test seeks to measure. The questionnaire was evaluated by subject matter experts for content validity. The subject matter experts evaluated the questionnaire to ensure that (1) all the relevant technologies are listed, (2) all the relevant applications of the technology are listed and (3) the layout is logical and all relevant questions, required to collect data to answer the research questions, were included. The subject matter experts were information system experts, SME experts in Soweto itself as well as a questionnaire development expert.

Internal validity is the extent to which the questionnaire allows the researchers to draw conclusions about the relationship between variables. Internal validity was tested through various correlations and cross-tabulations. External validity, on the other hand, is the extent to which the sample genuinely represents the population from which it was drawn. In this research the sample was representative as the respondents were all SME owners within Soweto.

The results from this survey can be used with confidence as the reliability and validity of the questionnaire were proven through the reliability and validity tests.

RESULTS AND ANALYSIS

The first section of the results focuses on the biographical information of the respondents. It provides an overview of the respondents with regard to their income and type of small business that they own and manage.

The results from Figure 3 indicate that the majority (79.2%) of the SME owners earned less than R5000 a month.

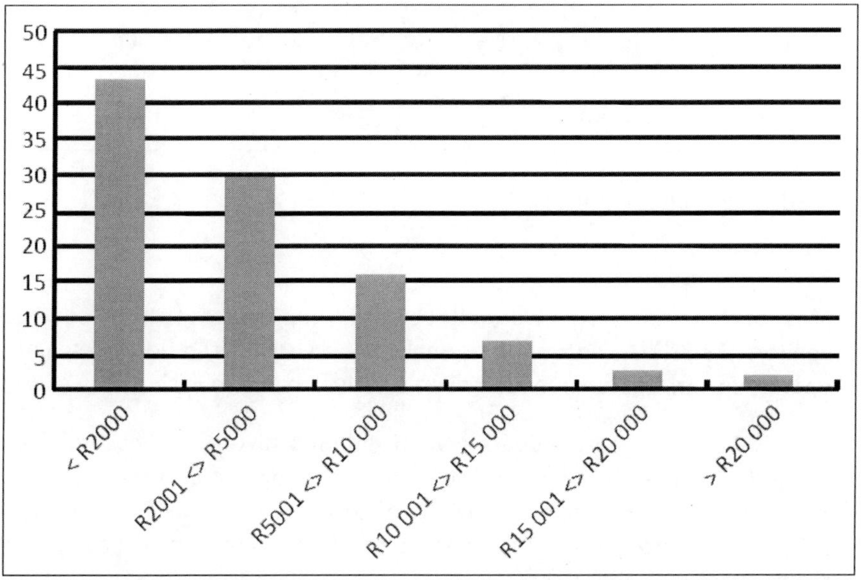

FIGURE 3: Monthly income.

What is even more astounding is that 43.3% of the respondents earned less than R2000 a month through their business. These results indicate then that close to 80% of the MSBs in this survey can be classified as micro enterprises which include survivalist MSBs.

Only 4.4% of the respondents earned more than R15 000 per month.

The next question on the questionnaire focused on the type of business that the respondents owned. These results are displayed in Figure 4.

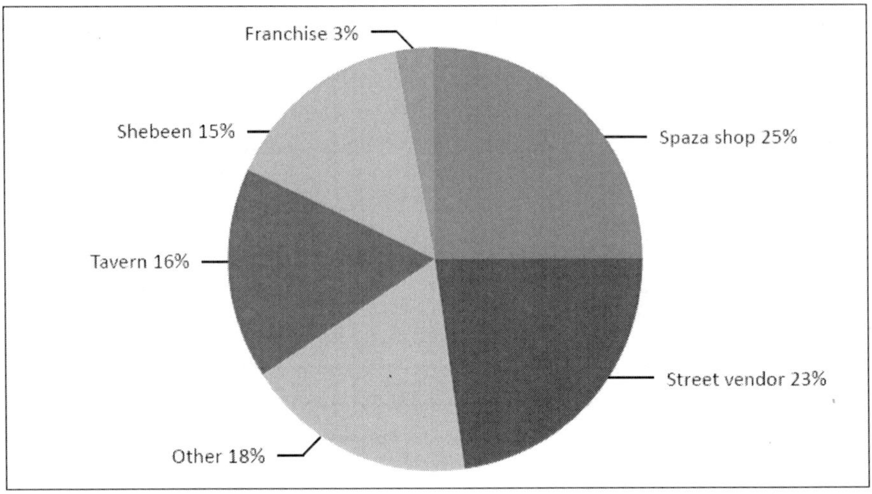

FIGURE 4: Type of business.

Most of the respondents were either *spaza* shop owners (24.8%) or street vendors (23.3%). A *spaza* shop normally has a very limited floor area and is operated from a garage, shed or even a bedroom in a township.

Some 30% of the respondents owned either a tavern or *shebeen* (places that sell alcohol). Only 3.1% of the respondents owned a formal franchise. The 'Other' category made up 18.3% of the remainder of the businesses. MSBs in this grouping include barber shops, hairdressers, daycare, shoe repair and public phone rentals.

Table 4 presents a cross-analysis of the income versus the type of business. It is evident from Table 4 that the more formal the type of MSB or SME, the more income is generated.

TABLE 4: Cross-analysis of income and type of business.

Type of business	< R2000 (%)	R2001 <> R5000 (%)	R5001 <> R10 000 (%)	R10 001 <> R15 000 (%)	R15 001 <> R20 000 (%)	> R20 000 (%)
Street vendor	19.0	4.0	0.3	-	0.1	-
Spaza shop	10.0	8.0	5.0	1.0	0.1	0.2
Franchise	1.0	0.4	0.2	1.0	1.0	0.3
Shebeen	4.0	7.0	3.0	1.0	0.3	0.1
Tavern	1.0	6.0	4.0	3.0	1.0	1.0
Other	8.0	5.0	3.0	1.0	1.0	1.0
TOTAL (%)	**43.00**	**30.40**	**15.50**	**7.00**	**3.50**	**2.60**

Street vendors and *spaza* shops accounted for 29% of the income, which was less than R2000 per month.

The results from the biographical data highlight that most of the MSBs can be classified as micro enterprises, where the income was less than R150 000 per year. Only 13.1% can be classified as small MSBs.

The second part of the analysis focuses on the type of ICT used by the owners of the MSBs, the frequency of the usage as well as the reasons for using the specific technology.

Figure 5 displays the type of ICT used and the frequency of usage. It must be noted that the respondents could select one or more of the listed technologies. The idea was to determine how many technologies were used as well as their frequency of usage.

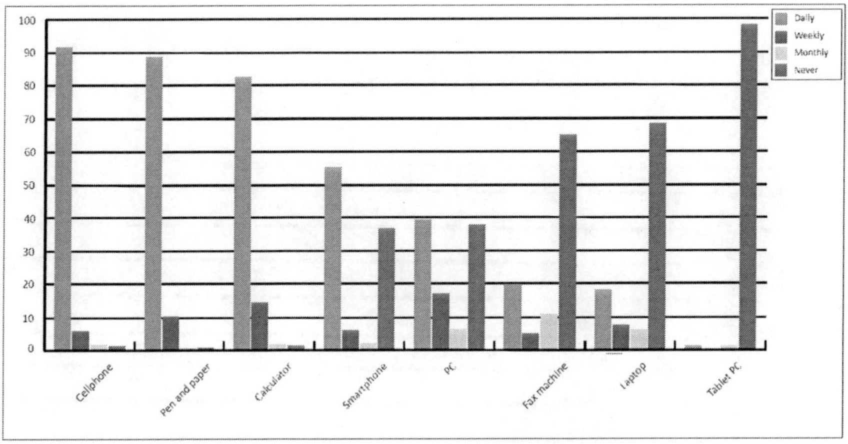

FIGURE 5: Frequency and usage of information and communications technology

The results indicate that the 'traditional' technologies were still used most of the time. Normal basic cellphones, without any features, as well as pen and paper were the technologies that were used the most. They were also used on a constant, daily basis. The technology that also featured fairly often (82.6% daily usage) was calculators.

Although statistics indicate that smartphones are infiltrating the cellular market, only 55.3% of the respondents used them. It is quite peculiar that both cellphones and smartphones were used by the respondents, highlighting the fact that some South Africans have more than one mobile device per person.

On the other side of the scale are the technologies that were not used. These include tablet devices (98.1% never used), laptops (68.4% never used) and fax machines (65% never used). Although these technologies were mostly not used, it must be noted that where they were indeed used, fax machines (19.6%) and laptops (18%) were actually used on a daily basis.

The following section provides an analysis of the technologies to determine the purpose for which the specific technology was used.

Type of ICT and usage

The results indicate that cellphones were used for what they were intended. The respondents used them for normal day-to-day phone calls (83.2%) as well as to send SMSs (61.6%). The results are depicted in Figure 6.

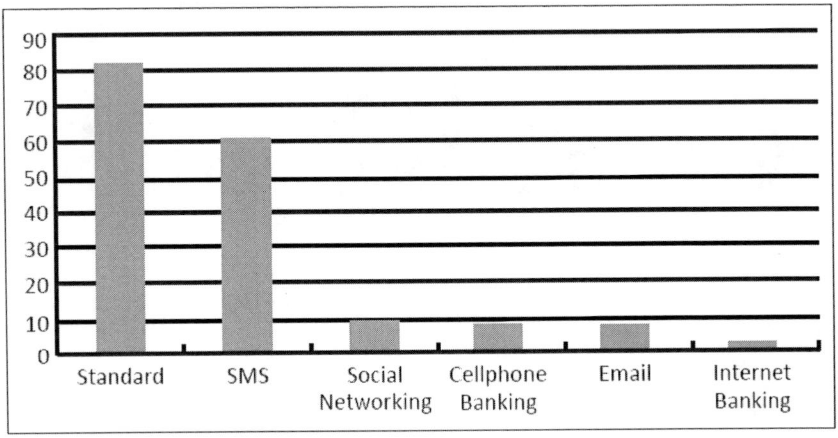

FIGURE 6: Cellphone usage.

It is of interest that only a small portion of the respondents used cellphones for business purposes, such as cellphone banking (8%), email (7.5%) and Internet banking (2.4%).

Although the respondents did not use smartphones as much as cellphones, it seems that those who did use them did so more for business-related activities. This indicates a paradigm shift as depicted in Figure 7.

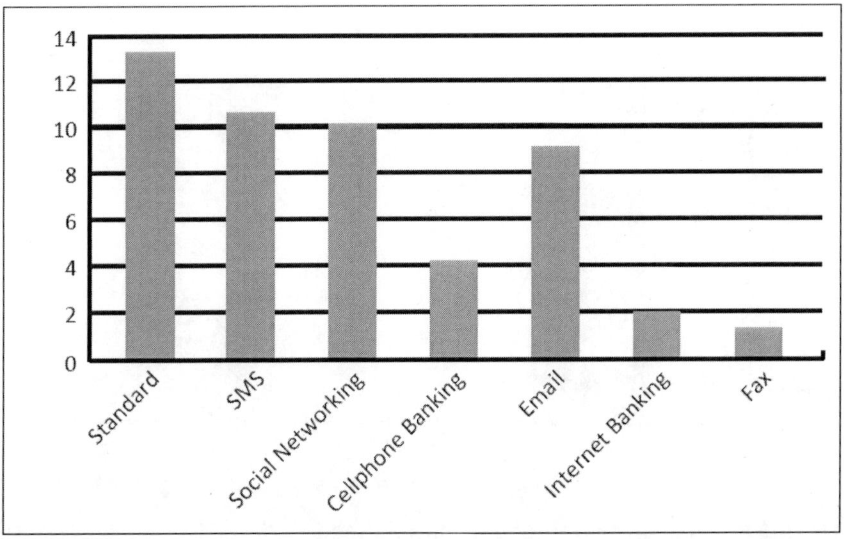

FIGURE 7: Smartphone usage.

The results must be viewed in relation to the fact that only 55% of the respondents used a smartphone on a daily basis. The implication is that those respondents who had two phones (a cellphone and smartphone) used the cellphone for normal usage and the smartphone for business-related activities.

When the usage of PCs and laptops are compared, no significant difference in the adoption and usage of the technology can be seen. A chi-squared test was done to test for independence between the usage of PCs and laptops and the results indicated that there is no dependency between the two variables. The 100% stacked bar in Figure 8 highlights these results.

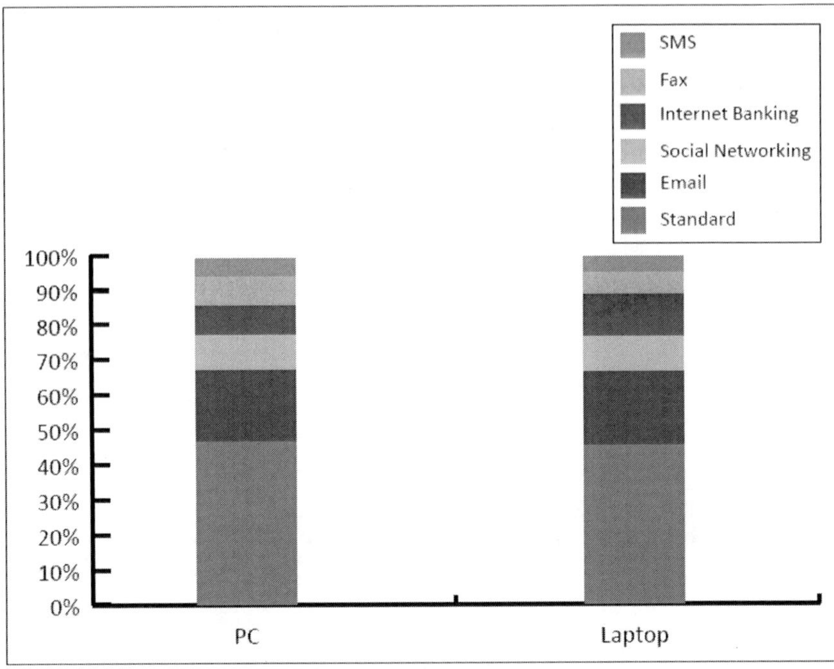

FIGURE 8: Personal computer and laptop usage.

The results indicate that those respondents who used a PC on a daily basis (39.2%) did not necessarily use it for business-related activities. The usage of laptops paints a similar picture and it is evident that laptops were not significantly used by the respondents. One might speculate that the high crime rate in this area could play a role in this statistic, as it is easy to steal a laptop.

The results, as depicted in Figure 6, clearly indicate that there is no difference between the usage of PCs and laptops. Each of the applications is, per ratio, the same with each technology.

Figure 9 also paints an interesting picture. The Internet itself, when it was available, was used for normal browsing (47%) and social networking (19%). It was seldom used for Internet banking (10%) and email (24%).

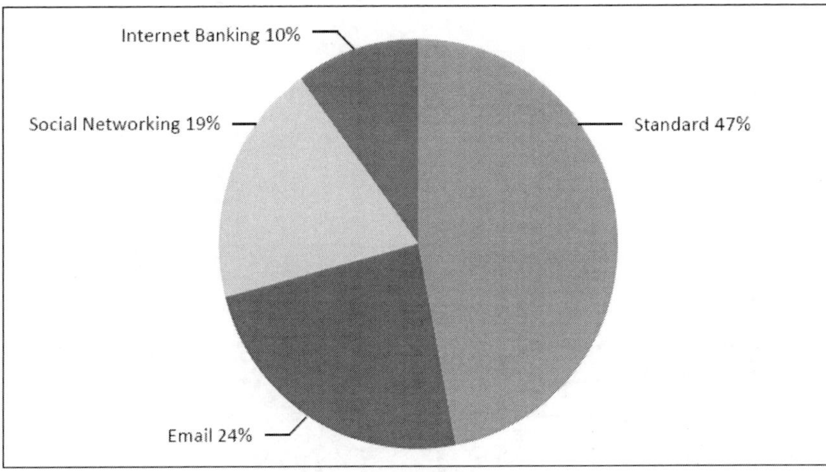

FIGURE 9: Proportional Internet usage.

This might result from limited Internet access in South Africa, as shown in Figure 2.

The next section investigates the correlations to determine whether or not more formal MSBs are adopting ICT and the purposes of its use.

The technique of multiple responses was used to undertake further analysis. This technique allows the analysis of a number of separate variables at the same time. According to Argyrous (2011:512): 'this is used in situations where the responses of separate variables that have a similar coding scheme all point to the same underlying variable'. In this case, the underlying variables were the four major types of technologies: cellphone, smartphone, PC and laptop. Frequencies and cross-tabulations can be performed with multiple responses.

Figure 10 shows the correlation between the type of ICT usage and the type of SME. The purpose of this correlation was to determine whether or not the more formal MSBs used more high-end technology, such as PCs and laptops.

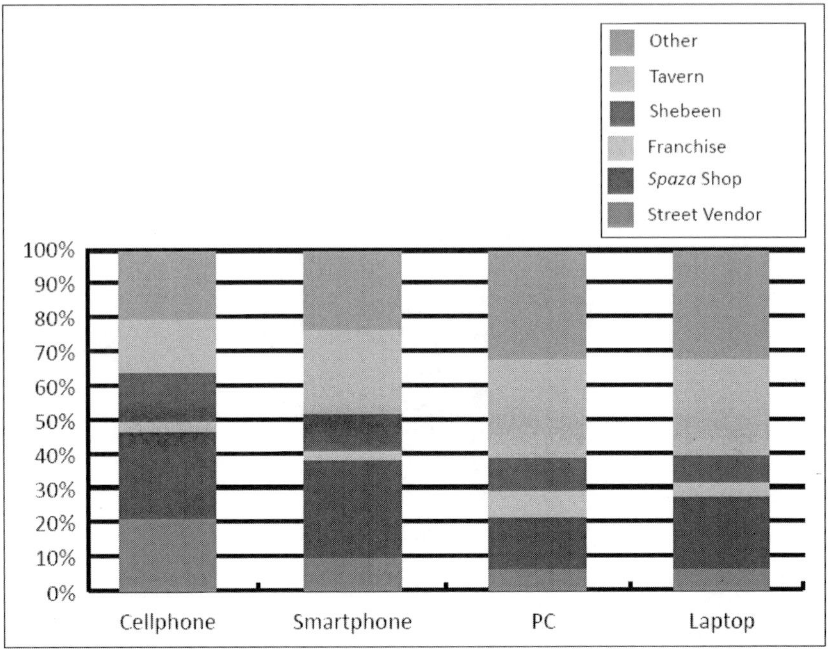

FIGURE 10: Correlation between information and communications technology usage and type of business.

The results in Figure 10 indicate that the more formal MSBs used more high-end technology than the informal MSBs. Cellphones were used by 47% of street vendors and *spaza* shop owners. This is in contrast with MSBs such as taverns (15.1%), which used cellphones in their day-to-day business operations.

PCs and laptops were, on the other hand, used by more formal MSBs (60.6% and 59.5%, respectively). Street vendors and *spaza* shops used PCs (22.2%) and laptops (27.7%) for business.

Further analysis of the results indicates that the MSBs, which were categorised within the 'Other' group, adopted the usage of PCs and laptops more than the other types of business. It is important to note at this point that 44.1% of the 'Other' consists of four types of MSBs: hair salons (21.8%), fast food outlets (10.1%), barbershops (6.1%) and Internet cafés (6.1%).

Figure 11 highlights that the other types of MSBs used PCs more for business-related activities than the rest of the MSBs. This might be related to the type of business, for example, fast food outlets might need a PC to transact daily sales.

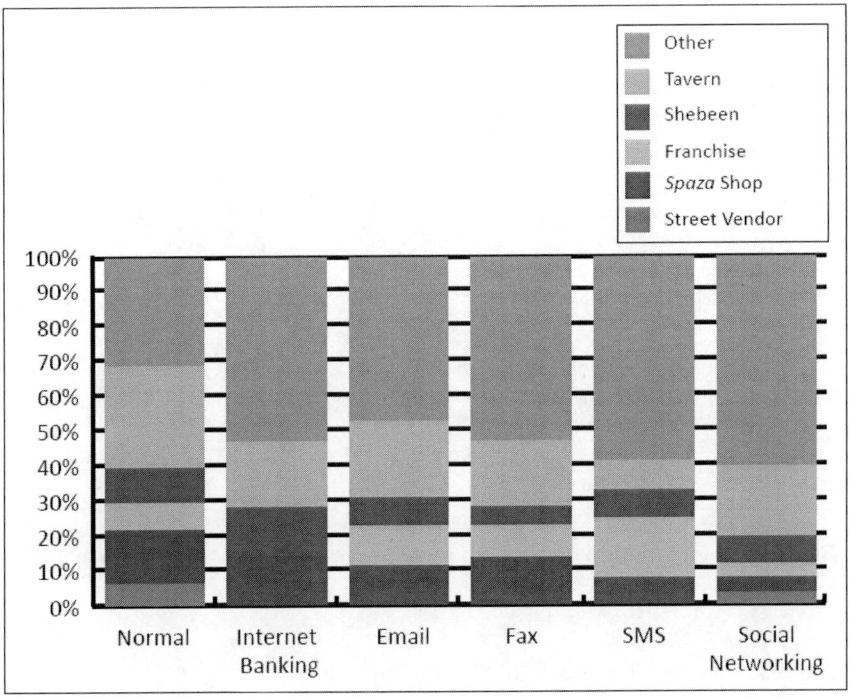

FIGURE 11: Correlation between personal computer usage and type of business.

Figure 12 tells the same story as with PC usage, where the other types of MSBs used laptops more frequently for business than, for instance, franchises. Table 5 is a comparison between the adoption of cellphones versus smartphones. The comparison is done per type of business.

TABLE 5: Cellphone versus smartphone usage.

Type of business	Normal (%)		Internet banking (%)		Cellphone banking (%)		Email (%)		Fax (%)	SMS (%)		Social networking (%)	
	Smartphone	Cellphone	Smartphone	Cellphone	Smartphone	Cellphone	Smartphone	Cellphone	Smartphone	Smartphone	Cellphone	Smartphone	Cellphone
Street vendor	10	21.9	5	4.3	7.1	11.5	4.4	11	7.1	8.6	20.8	9	20.9
Spaza shop	26.9	25.4	20	21.7	11.9	26.9	25.6	21.9	14.3	26.7	24.1	24	17.6
Franchise	3.1	3.2	5	0	2.4	1.3	2.2	4.2	7.1	2.9	5.2	4	1.1
Shebeen	11.5	14.7	5	17.4	9.5	17.9	11.1	12.3	14.3	11.4	15.1	12	7.7
Tavern	26.5	15	30	34.8	26.2	21.8	33.3	24.7	21.4	27.6	15.1	24	13.4
Other	21.5	19.8	35	21.7	42.9	20.5	23.3	26	35.7	22.9	21.8	27	37.4

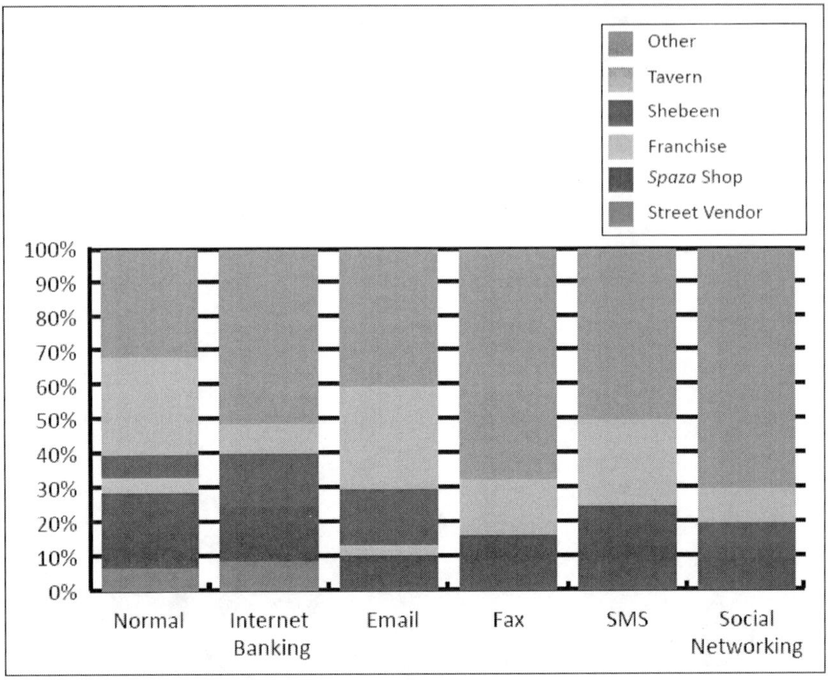

FIGURE 12: Correlation between laptop usage and type of business

The results as per Table 5 highlight the two major differences:

1. Cellphones were the preferred technology when the MSB used them for normal usage, such as phoning and sending SMSs. Smartphones, on the other hand, were used when the focus was on business-related transactions, such as Internet banking and emails.

2. The more formal MSBs used smartphones instead of cellphones. Street vendors and *spaza* shops were more inclined to use cellphones.

The second correlation focuses on the income and the adoption of ICT. As per the literature, the assumption is that the adoption of high-end technology will increase the profits and turnover of the SME. Figure 13 shows the correlation between the monthly income and the type of technology used.

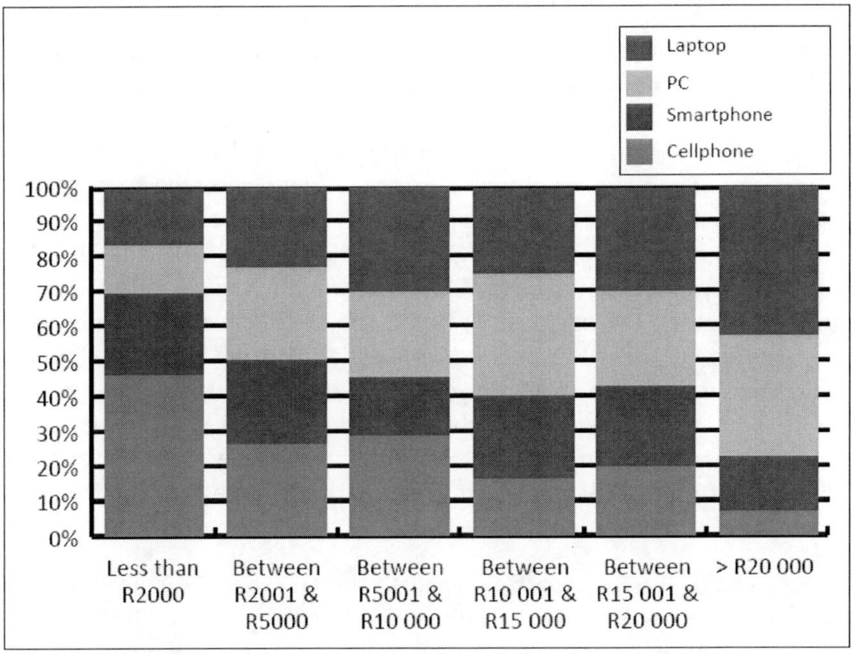

FIGURE 13: Monthly income versus type of technology

The results clearly highlight this positive correlation between the monthly income and the type of technology adopted by the SME. The monthly income was higher where PCs and laptops were used in the day-to-day running of the SME. This raises an interesting conundrum: must MSBs invest in high-end technology hoping that it will generate money, or is it an evolving concept?

An in-depth analysis of the results indicates that a positive correlation exists between the usage of ICT and the income of the business. ICT is an enabler even in township MSBs.

DISCUSSION

MSBs play an important role within any economy and even more so within developing economies such as South Africa. The aim of this research was to provide statistical information on the 'raw' rates of ICT adoption amongst MSBs in Soweto. Four research questions were posed and are discussed individually.

The first question focused on the type of ICT adopted by the various MSBs. The results indicate firstly that traditional pen and paper as well as cellphones are the technologies that are mostly used. A calculator is also a technology that is used on a daily basis. The interesting fact is that only a small portion of the MSBs surveyed used PCs and laptops for the daily management and running of their business. Tablet devices were not used at all even when these devices were given to the businesses for free, with the renewal of cellphone contracts. The assumption can be made that the functionality and benefits of tablet devices are not known and realised by the SME owners. This might result from a lack of education. Tablets are also not very useful without a network connection. Tablets with SIM slots are quite expensive and the cheaper versions, with Wi-Fi only, are not useful as there is no Wi-Fi network available in Soweto. If free Wi-Fi is funded by government, private industry or both, there would be an increase in the adoption of these cheaper tablets. The Internet is an essential part in the process of technology adoption.

The results of the first question highlight the fact that ICT is not that well accepted and adopted. That leads to the second question which tries to uncover whether or not the adoption of ICT is related to the type of SME that may use it. The results confirm that the usage of ICT is directly linked to the type of SME and directly to the income of the business. The more formal the type of SME and, thus, the higher the income, the more the usage of advanced ICT such as PCs and laptops. MSBs in the 'Other' type of SME grouping indicate that these MSBs use ICT to the SME's benefit. ICT is used for a wide variety of functions from social media to banking and email.

The third question focused on the correlation between the type of ICT adopted and the profitability of the SME itself. The results clearly highlight this positive correlation between the monthly income and the type of ICT adopted by the SME. The monthly income is higher where PCs and laptops are used in the day-to-day running of the SME. What is not clear from the results is whether ICT boosts the business, or whether the more formal MSBs by nature adopt and use ICT. This requires further investigation about what came first: ICT that helped the SME or the SME that used ICT and, thus, grew even further.

The fourth question interrogated the relationship between applications and ICT based on the type of SME. In other words, is the cellphone applied differently between a street vendor and a franchise? No definitive conclusion can be made. What is clear is that smartphones are the preferred technology, over cellphones, for conducting business.

It is clear that the MSBs in Soweto at best make general use of ICTs which include email and Internet access. The rates of adoption are very high and do not depend on the size of the SME. If there was a sliding scale, the MSBs would have featured fairly low on it, even as general users of ICTs. That should be a national concern about why these MSBs are not adopting ICT.

The results are quite disheartening from an ICT perspective and they also provide additional information regarding the type and size of the MSBs in Soweto. ICT is particularly easy to use these days and, thus, the adoption of ICT must rely on the perceived usefulness. This can only be overcome through training and education. Sidney *et al.* (2010) and also Santosa and Kusumawardani (2010) emphasise that there must be an effective implementation strategy that focuses on the deployment of ICT for MSBs. The strategy should serve as a guideline for MSBs to improve they utilise their ICT infrastructure.

The analysis of the results also indicates that although ICT is used by some MSBs, the usage is not necessarily linked to business. ICT is used for its intended purpose, such as word processing, but it is not used or adopted for business requirements such as marketing, Internet banking and financial analysis. It might be a matter of education or ignorance, but this should be researched further.

The results indicate that most MSBs are either survivalists or very small MSBs that barely make a living. It makes logical sense that ICT does not necessarily feature within the framework of such MSBs. However, ICT can, as an enabler, contribute to growing the business and generating more income, and, thus, move the SME to the next level.

CONCLUSION

The literature highlights the importance of MSBs within developing economies and their vital role in the growth of the economy. The literature also highlights the importance of ICT as an enabler for business and the argument is, thus, that ICT can or should enable MSBs to grow themselves and, thus, stimulate the economy of a country such as South Africa.

The findings indicate that the adoption and use of ICT are directly linked to the income of the respective MSBs. The more formal MSBs use more advanced technologies such as PCs and laptops, whereas the smaller MSBs revert back to pen and paper, calculators and cellphones. Another aspect that was highlighted is the fact that ICT, when it is utilised, is not fully utilised. An example is cellphones which are used mostly for normal, non-business purposes. The SME owners have not yet made the transition from traditional usage of cellphones and the intent, to applying these for business usage and growing the business. A startling aspect is that most of the MSBs are survivalist MSBs, focusing as such more on surviving and not necessarily on growing the business.

It can be concluded that work remains to be completed concerning training on and insight into the usage of ICT. The results highlight the fact that SME owners do not necessarily know what they can do with ICT and how it can grow their business. A possible solution is to liaise with the local business chambers within Soweto and provide the necessary training and awareness as a continued effort and not just a once-off scenario.

It is the first time that such research has been performed within a South African township and the results are significant for future research. The results from this research open an entirely new world of research possibilities. Various areas of research never addressed before have also been highlighted.

Future research should include a more in-depth analysis of why ICT is not used for business, and what those various barriers are to entering the world of ICT. This was not covered by this research, as the aim was to establish a common understanding of the problem at hand. The technology adoption model (TAM) is a causal model and can, thus, be used to test certain hypotheses. Future research can also be expanded to other South African townships to determine the differences between the various populations.

For township MSBs to embrace ICT and use it to the fullest, the full potential of ICT must be advocated and the benefits must be worthwhile. If this is not achieved with a concerted effort, then the digital divide will not be crossed by MSBs and they will not benefit from the potential growth and market share. ICT might help most of the MSBs to move from survivalist to micro enterprises where they can start employing other township people and, thus, reduce the unemployment statistics of South Africa.

REFERENCES

1. Abor, J. & Quartey, P., 2010, 'Issues in SME Development in Ghana and South Africa', *International Research Journal of Finance & Economics* 39, 218–228.

2. Arendt, L., 2008, 'Barriers to ICT adoption in SMEs: How to bridge the digital divide?', *Journal of Systems and Information Technology* 10, 93–108.

3. Argyrous, G., 2011, *Statistics for research with a guide to SPSS*, Sage Publications, London.

4. Balnaves, M. & Caputi, P., 2001, *Introduction to quantitative research methods – An investigative approach*, Sage Publications, London.

5. Beck, T., Demirguc-Kunt, A. & Levine, R., 2005, 'SMEs, growth, and poverty: Cross-country evidence', *Journal of Economic Growth* 10(3), 199–229.

6. Blaikie, N., 2003, *Analyzing Quantitative Data – From Description to Explanation*, Sage Publications, London.

7. Blueprint Strategy & Policy (Pty) Ltd., 2005, *Promotion of small and medium enterprises in the South African chemicals sector*, Blueprint Strategy & Policy (Pty) Ltd., South Africa.

8. Cameron, S. & Price, D., 2009, *Business research methods – A practical approach*, Chartered Institute of Personnel and Development, London.

9. Cravo, T., Gourlay, A. & Becker, B., 2012, 'SMEs and regional economic growth in Brazil', *Small Business Economics* 38(2), 217–230.

10. Cronbach, L., 1951, 'Coefficient alpha and the internal structure of tests', *Psychometrika* 16(3), 297–334.

11. Davis, F.D., 1989, 'Perceived Usefulness, Perceived Ease of Use, and User Acceptance of Information Technology', *MIS Quarterly* 13(3), 319–340.

12. Davis, F.D., Bagozzi, R.P. & Warshaw, P.R., 1989, 'User acceptance of computer technology: A comparison of two theoretical models', *Management Science* 35(8), 982–1003.

13. European Commission, 2012, 'Small and medium-sized enterprises (SMEs): What is an SME?', *European Commission* , viewed 04 June 2012, from http://ec.europa.eu/enterprise/policies/sme/facts-figures-analysis/sme-definition/index_en.htm

14. Falkena, H., n.d., 'SMES' access to finance in South Africa – A supply-side regulatory review', in H. Falkena, (ed.), *Financial Services and Regulation*, South Africa.

15. Feilzer, Y.M., 2010, 'Doing mixed methods research pragmatically: Implications for the rediscovery of pragmatism as a research paradigm', *Journal of Mixed Methods Research* 4(1), 6–16.

16. Hotho, S. & Champion, K., 2011, 'Small businesses in the new creative industries: Innovation as a people management challenge', *Management Decision* 49(1), 29–54.

17. Hu, Q. & Plant, R., 2001, 'An empirical study of the casual relationship between IT investment and firm performance',*Information Resources Management Journal* 14(3), 15–26.

18. Huizingh, E., 2007, *Applied statistics with SPSS*, Sage Publications, London.

19. Kannabiran, G. & Dharmalingam, P., 2012, 'Enablers and inhibitors of advanced information technologies adoption by SMEs: An empirical study of auto ancillaries in India', *Journal of Enterprise Information Management* 25(2), 186–209.

20. Kennedy, C., Bounds, M. & Goldman, G., 2011, 'Strategic planning in SMEs: conscious effort or afterthought?', in G. Goldman & G. Heyns (eds.), *The 4th International UJ Faculty of Management Conference*, 29–31 May 2012, Faculty of Management, University of Johannesburg, Amanzingwe Lodge, Broederstroom, Gauteng, South Africa.

21. Khalifa, M. & Davison, M., 2006, 'SME adoption of IT: the case of electronic trading systems', *IEEE Transactions on Engineering Management* 53(2), 275–284.

22. Kirk, J. & Miller, M.L., 1986, *Reliability and validity in qualitative research*, Sage Publications, Beverly Hills.

23. Kitchenham, B. & Pfleeger, S.L., 2002, 'Principles of survey research part 4: Questionnaire evaluation', *SIGSOFT Softw. Eng. Notes* 27(3), 20–23.

24. Litwin, M.S., 1995, *How to measure survey reliability and validity*, Sage Publications, London.

25. Lucchetti, R. & Sterlacchini, A., 2004, 'The adoption of ICT among SMEs: Evidence from an Italian survey', *Small Business Economics* 23(2), 151–168.

26. Mahembe, E., 2011, *Literature review on small and medium enterprises' access to credit and support in South Africa*, National Credit Regulator, South Africa.

27. Mbonyane, B.L., 2006, *An exploration of factors that lead to failure of small businesses in the Kagiso*, M.Tech Business Administration, UNISA, Pretoria.

28. Mitra, S., 2005, 'Information Technology as an Enabler of Growth in Firms: An Empirical Assessment', *Journal of Management Information Systems* 22(2), 279–300.

29. Ngassam, E.K., Ntawanga, F. & Eloff, J.H.P., 2013, 'A roadmap for rural area ICT solution deployment: A Case of Kgautswane community in South Africa', *The African Journal of Information Systems* 5(2), 9–64.

30. Njiro, E., Mazwai, T. & Urban, B., 2010, 'A situational analysis of small businesses and enterprises in the townships of the Gauteng province of South Africa,' in 2010 *Soweto International Conference on Entrepreneurship & Development*, Soweto, Johannesburg.

31. Olawale, F. & Garwe, D., 2010, 'Obstacles to the growth of new SMEs in South Africa: A principal component analysis approach', *African Journal of Business Management* 4(5), 729–738.

32. Polun, C., Yueh-Shuang, H., Yuann-Meei, T., Yiing-Yiing, S., Hou, I.C. & Wei-Fong, K., 2004, 'The development of intelligent, triage-based, mass-gathering emergency medical service PDA support systems', *Journal of Nursing Research (Taiwan Nurses Association)* 12(3), 227–235.

33. Premkumar, G. & Roberts, M., 1999, 'Adoption of new information technologies in rural small businesses', *Omega* 27(4), 467–484.

34. Santosa, P.I. & Kusumawardani, S.S., 2010, 'Improving SME ICT utilization through industrial attachment program: Indonesia case', in *Frontiers in Education Conference* (FIE), 2010 IEEE, 27–30 Oct. 2010, S1J–1–S1J–5.

35. SBP., 2013, *SME growth index 2012 headline report. easier/harder for small business in South Africa?* SBP, Johannesburg.

36. Sidney, K.T.T., Tina, A.B. & Choy, K.L., 2010, 'Efficacy of IT/IS deployment in small– and medium enterprise in China', in*2010 8th International Conference on Supply Chain Management and Information Systems (SCMIS)*, 6–9 October, 2010, Lisbon, Portugal, pp. 1–7.

37. Silvius, G., 2004, 'IT and small business: An unhappy marriage?', in M. Khosrow-Pour (ed.), *Innovations through information technology*, Idea Group Publishing, Hershey.

38. Statistics South Africa, 2012, *Census 2011 statistical release*, Statistics South Africa, Pretoria.

39. Terblanche, N.S., 1991, 'The spaza shop: South Africa's first own black retailing institution', *International Journal of Retail & Distribution Management* 19(5), 37–43.

40. The World Bank, 2006, *Information and Communications for Development – Global Trends and Policies*, The World Bank, Washington D.C.

41. Thong, J.Y.L. & Yap, C.S., 1995, 'CEO characteristics, organizational characteristics and information technology adoption in small businesses', *Omega* 23(4), 429–442.

42. Thurik, R. & Wennekers, S., 2004, 'Entrepreneurship, small business and economic growth', *Journal of Small Business and Enterprise Development* 11(1), 140–149.

43. Venkatesh, V., Morris, M.G., Davis, G.B. & Davis, F.D., 2003, 'User acceptance of information technology: Toward a unified view', *MIS Quarterly* 27(3), 425–478.

44. Yonazi, E., Kelly, T., Halewood, N. & Blackman, C., 2012, *The transformational use of information and communication technologies in Africa*, eTransfrom Africa.

Chapter 11

Convergence of Functional Areas in Information Operations

Brett van Niekerk[1]

[1]School of Management, IT and Governance, University of KwaZulu-Natal, South Africa

ABSTRACT

Background: Contemporary technology and its rapid rise to a ubiquitous nature are affecting the way in which technology is used. This holds implications for military information operations and information warfare concepts as the previously separate functional areas are increasingly overlapping due to the increased convergence of information technology. Objective: The aim of the article is to illustrate the convergence of the functional areas of information operations through the use of reported examples and descriptive models.

Methods: A qualitative review and analysis of practitioner documents, academic publications, and other reports is conducted. The diverse views of this phenomenon are contrasted and discussed. The possible implications of the convergence and possible management techniques are proposed and discussed.

Results: There is strong opinion that the domains are not converging, however practice shows that there is an increasing overlap of operational considerations and organisational structure.

Conclusion: It is concluded that the domains are not converging, but the operations and management of the information operations functional areas may converge, and this has an impact across all functional areas that needs to be taken into consideration.

INTRODUCTION

Contemporary technology such as social media, the rapidly growing prevalence of wireless networking technology and the expanding capabilities of smart mobile devices result in what is termed as digital convergence. This has an impact on the military use of information technology, namely information operations (IO) and information warfare (IW), as the technology upon which the different functional areas are based converge and evolve. Through the use of examples and descriptive models, the convergence of the functional areas of information warfare and information operations is discussed.

An overview of information operations, information warfare and their constituent functional areas is provided, after which the existing relationship amongst the main functional areas is described through the use of generally accepted models of information technology and reviews of existing debates on the topic. The convergence through the prevalence of contemporary information technology is illustrated through the use of reported examples, and the implications of the convergence and management thereof are discussed.

METHODOLOGY

The convergence of information operations, in particular between electronic warfare and computer network operations, is being debated, mostly in practitioner forums. There is limited academic contribution to these debates, but it has been indicated that there is a need for academic models for information operations (Armistead 2010:108). This article aims to fill the gap by providing an academic perspective on the debate and by proposing an alternative model to those discussed previously.

The research takes an interpretivist stance in that the analysed documents that consider convergence in information operations are often authored by practitioners whose reasoning is not only intended to understand the concept but to actively participate in defining it. The approach taken is inductive, seeking to develop theory on the convergence amongst the functional areas in information operations. A qualitative document analysis of reports and examples, combined with logical argument, is used to derive on-going convergence and the potential impact thereof. Rouse and Dick (1994:51) illustrate the importance of qualitative research for information systems '... to capture holistic real-world answers to real-world problems in a way that is not possible in a quantitative context'. Publication databases and search engines were used to identify academic and practitioner publications and documents which consider the topics of interest. Search engines were used to identify relevant news reports that illustrate the existence of the concepts under discussion. A total of 11 academic or practitioner publications and 18 news reports were identified. All of these are considered in the article.

Functional areas in information operations

This section describes the various functional areas of information operations. The definition of IO and IW varies by country and organisation. Usually this reflects the specific strategy or focus that is employed. For this paper, two definitions will be considered. Denning (1999) defines IW as the following:

... offensive and defensive operations against information resources of a 'win-lose' nature. It is conducted because information resources have value to people. Offensive operations aim to increase this value for the offence while decreasing it for the defence. Defensive operations seek to counter potential losses in value. (p. 21)

Brazzoli (2007) considers IW to be as follows:

All actions taken to defend the military's information-based processes, information systems and communications networks and to destroy, neutralise or exploit the enemy's similar capabilities within the physical, information and cognitive domains. (p. 219)

These definitions highlight the value of information and that certain activities seek to change or maintain the current value or status of the information through activities relating to information assets and related infrastructure and systems. The notion of different domains is also raised: people or psychological, virtual and physical aspects of information. Whilst the military aspect is emphasised, Cronin and Crawford (1999) indicate that information operations can be applied to non-military situations, supporting the views considering information operations as an extension of

IW to non-conflict scenarios (such as peacetime and periods of increased competition).

Computer network operations (also known as cyber-warfare) focus on computer networks and the Internet. It comprises of computer network exploitation (intelligence gathering), computer network attack, computer network defence and computer network support. In some instances, this is extended to include the protection of critical infrastructure from attacks through computer networks (Brazzoli 2007:221; Van Niekerk 2011:39–40). This functional area is the newest functional area and is yet to be fully established.

Electronic warfare (EW) focuses on the electromagnetic spectrum, comprising of electronic protection, electronic support and electronic attack. This most commonly includes the jamming, detection and interception of signals (Joint Chiefs of Staff 1998:II–5).

Psychological operations (PSYOP) are actions taken to influence a population's perceptions, attitudes and behaviour (Brazzoli 2007:221). PSYOP can be considered to be limited to military operational areas whereas strategic communication and propaganda can be considered to be of a larger scale, aimed at national or international audiences ranging from competition to conflict scenarios.

Other notable areas deal with intelligence gathering, command and control and the network support to operations and decision-making. It also includes data and audio and imagery communication via wired and wireless networks and infrastructure (Brazzoli 2007:221). This results in a shift from the traditional platform-centric operations and command and control to the network-centric form where information can be shared amongst platforms so that everyone has access to the same information. The intelligence process of gathering, analysing and disseminating data and information to support the decision-making process (command and control) is therefore key. Offensive operations attempt to disrupt these capabilities of the adversary, and defensive operations seek to maintain these capabilities in allied and own forces (Brazzoli 2007:221).

Porche *et al.* (2013:25–28) indicate that there are problems with the current definition and construction of IO: There is a lack of common vision ranging from too focussed to too broad. Many functional areas are treated as compartmentalised (Porche *et al.* 2013:24), and there is no explanation as to how it will act as a co-ordinating function or as to the relationships amongst the different functional areas. The next section describes the relationships between different functional areas of IO through the use of examples and positions advocated in documents.

Relationships amongst functional areas

There has been much debate regarding the relationship between EW, cyber-warfare and computer network operations and the implications of the various forms of terminology. Smith and Knight (2005:53) apply EW concepts to network security. This indicates a number of parallels between EW and cyber-warfare such as jamming and denial-of-service (DoS) or flares and honeypots.

A report by the US Government Accountability Office indicates that the two domains should be considered as separate (Chabrow 2012). Knowles (2013:48) proposes that cyber-space is a man-made physical environment, including telecommunication, data, networks and processors. In contrast, the electromagnetic spectrum is natural. Wireless communication is seen as the human exploitation of an existing natural space whereas cyber-space is a purely man-made construct. However, it can be argued that the physical components of the telecommunication and data networks that form cyber-space are based upon guided or radiated electromagnetic signals and therefore part of the electromagnetic spectrum (Clifford 2011:41; Hahn 2010: 44–46). Clifford (2011:41) argues that it is incorrect to say that cyber-space and the electromagnetic spectrum are equivalent due to the fact that they are based on similar physics. Some of the discussion revolves around terminology: Borque (2008:30–40) discusses whether cyber-operations are a combination of EW and computer network operations and concludes that EW and cyber-operations are separate. Cloud (2007:10–12) takes the contradictory view that cyber-operations entail a broader domain and include the electromagnetic spectrum, indicating that EW is a subset of cyber-operations.

The use of the psychological and cyber-domains is also related. The concept of hacktivism can be considered a combination of two areas: Activism can be seen as a form of psychological operation, and hactivism combines it with cyber-warfare concepts. Therefore website defacement can be described as cyber-enabled PSYOP. In some instances, the cyber-attack is meant to create the psychological impact such as the tactics used by the group Anonymous. The attack on the website of the South African Police Service was a cyber-attack with the intent to cause a psychological impact. It is reported that sensitive information on whistle-blowers was publicly released and that the attackers claimed that it was in retaliation for slow investigations in the Marikana mine shooting (Tubbs 2013). Anonymous Africa targeted a number of websites in Zimbabwe and South Africa with DoS attacks in protest against the Zimbabwean Government (Alfreds 2013; Daily News Correspondent 2013). There are reports of Israel hacking into phone voicemail systems to leave messages (StrategyPage.com 2009). This can be seen as an example of cyber-delivered PSYOP. There is also a possible relationship with EW and PSYOP in that radio or television broadcasts could be jammed to prevent them from being used to incite violence as happened in Rwanda (Van Niekerk & Maharaj 2009:6).

A number of espionage operations have been conducted over the Internet. These are cyber-attacks based on malicious, code-infecting targeted systems which then copy files onto a server from where the attackers can retrieve them. Major attacks of this type are the GhostNet cyber-espionage in 2009 and Red October, discovered in 2013 after an estimated five years of operation (Higgins 2013; Information Warfare Monitor2009). In 2013, details surfaced of data-collection activities by US intelligence agencies that were monitoring telecommunications meta-data and online communications (Greenwald & MacAskill 2013). However the most recognised case is that of Wikileaks, a website which posts information from whistle-blowers online. The major releases which created both support and condemnation of the website internationally were the series of releases of US military information and diplomatic cables (Gragido & Pirc2011:193–195; Van Niekerk & Maharaj 2011:7–8). These incidents indicate a strong relationship between intelligence, counter-intelligence and cyber-space.

As EW targets primarily radiate the use of the EMS (but can conceivably have an effect on guided transmissions as directed-energy) and are concerned with the modulation and frequency of these transmissions, it is applicable to the physical layer of the OSI model and may extend to the data-link layer. As cyber-space is the interconnection of systems and networking, it can be considered to extend from the data-link layer to the presentation layer, and in some cases, it overlaps with the application layer. Other areas, such as PSYOP and command and control, occur at the application layer, as these are ultimately end-user functions where the key aspect is their presentation of information. The functional areas can similarly be mapped to the Internet Protocol (IP) layers as illustrated in Box 1.

BOX 1: Relationship of IW Areas to the OSI Model and IP Layers.

OSI Layers	IP Layers	Functional Area
Application		PSYOPs, C2W
Presentation	Application	
Session		Network Warfare
Transport	Transport	
Network	Network	
Data-link	Data-link	
Physical	Physical	Electronic Warfare

Maasdorp and du Plessis (2012:29–33) propose another layered model to describe the relationship amongst the various aspects of IW and IO. This is shown in Table 1. Three layers are proposed: the utility layer, which is what humans will work with and consciously use; the connection layer, which is the logical connections of a network; and the access layer, which is the electromagnetic transport of analogue or digital bits. Overlap between the layers and their corresponding IW functional areas are allowed through the grey areas (Maasdorp & du Plessis 2012:29–33).

TABLE 1: Proposed Layer Model for IW.

Layer	Functional area
Utility	C2W, PSYOPs
Grey area	
Connection	Network warfare
Grey area	
Access (wired and wireless)	Electronic warfare

The convergence of functional areas through contemporary information technology

Contemporary technology such as the ubiquitous nature of mobile and wireless technology and social media are increasing the overlap amongst the domains and may therefore result in the convergence of IW functional areas.

The most discussed area for potential convergence is cyber-operations and EW, largely due to the evolution of wireless and mobile technology. Retired General Cartwright expressed views regarding the links between EW and cyber-operations (Freedberg 2013b), which are corroborated by reports that the development of new EW technology for the US Army is being complicated by the role of cyber-warfare and new attack methods. Chabrow (2012) also indicates that the management of EW is being complicated by the role and synergies with computer network operations. In conjunction with the development of the new EW systems, a new doctrine for cyber-electromagnetic operations is being developed (Freedberg 2013a), and a new training centre for the US Army is to incorporate aspects of cyber-operations, EW and other communications-related actions (Gould 2013). This culminated in a US Army doctrine for cyber-electromagnetic activities where such operations are seen as the overlap amongst cyber-space operations, electronic warfare and spectrum management operations (Department of the Army 2014:1–2). Similarly, the US Air Force is developing systems for 'spectrum warfare' which will cover EW, cyber-operations and other aspects. Some of the associated programs include research into 'net-enabled electronic warfare technologies' (Keller 2013). This indicates that the electromagnetic spectrum, traditionally the focus of EW, is now being related to computer networks. Another project is aimed at detecting cyber-attacks through the use of radio-frequency measurement and signals intelligence (Prince 2012). This can be seen as an intersection of intelligence, EW and cyber-operations.

Mobile technology, in particular smartphones and tablets, has integrated a number of forms of technology, which would previously have been separate tools for conducting espionage. They have integrated navigation, camera and video capture and the ability to transmit these wirelessly. They support Web browsing, have integrated social media connectivity and can function as a basic telephone in addition to having a variety of other applications. These include a hacking tool and a method for PSYOP message delivery (Van Niekerk & Maharaj 2012:4–7).

The mobile-phone infrastructure is different from traditional fixed-line communications where the data and voice channels are separate. With mobile phones, it is feasible that data-based attacks can impact the voice channels (Amoroso 2013:102). In addition, by targeting the wireless transmission using EW, both voice and data can be simultaneously disrupted or compromised. The introduction of voice over IP (VOIP) allows voice to be carried over the data networks. Many instant-messaging applications provide support for text, voice, video or a combination of these, which are transmitted over data networks. These forms of communication are susceptible to standard network attacks. Examples may include disruption by DoS and possible interception of and eavesdropping on communication.

Social media is a useful tool for PSYOP, and as it is cyber-based, it can be seen as a combination of these two as cyber-security concerns to protect the accounts used in such operations will also be relevant. The open nature of social media enables operators to gather intelligence on targeted individuals or groups to enable them conduct PSYOP on these targets (Van Niekerk 2012). Social media can be used on a large-scale in what can be called social IW by allowing groups to protest online with global support and organise and guide physical protests. Examples of this are the Arab Spring demonstrations and the related Occupy movement (Kamzi 2011; Madrigal 2011).

A number of cases discussed in this section will commonly be called convergence. However, Knowles (2013:48) suggests that convergence as a term is over-used and often misused and that sharing is a more appropriate term when discussing this phenomenon. The motivations for separate domains discussed in this section are largely limited to the domains. It is possible that the domains themselves remain separate, but the operations and management related to those domains converge. Therefore, the functional areas of IO can converge. Maasdorp and du Plessis (2012:32–33) support this view to an extent, indicating that there should at least be interaction between researchers and operators in the various fields to increase the efficiency and effectiveness of efforts. The increase in combined training for electronic warfare and cyber-operations indicates that there is a strong view that operations, or the management thereof, in these domains can be combined or integrated from a practical perspective.

Summary of convergence concepts regarding information operations

This section summarises the prior discussions regarding convergence. Table 2 illustrates the number of discussions related to the convergence between specific functional areas. As is evident, the focus is on cyber-operations and the electromagnetic spectrum. There is also some focus on the psychological aspects of cyber-operations, probably due to the number of hacktivist or propaganda-motivated cyber-attacks. The weakest overlap is between EW and PSYOP. Whilst there is some possible overlap between this two, the scope is very limited.

TABLE 2: Mentions of areas converging.

Cyber-EW	Cyber-intelligence	Cyber-PSYOP	EW-PSYOP
12	4	7	1

Table 3 illustrates the sentiment regarding convergence: Documents that take a pro-convergence perspective slightly outnumber those that take an anti-convergence perspective.

TABLE 3: Sentiment regarding convergence.

Pro-convergence	Anti-convergence
6	4

Implications of the convergence of functional areas in information operations

This section discusses the implications of the convergence of the functional areas of IO. These affects could be ethical, legal and operational and could affect the structuring of units or organisations and require specific management techniques.

The disruptive nature of networked communications

Whilst cyber-operations are still emerging and not yet fully established, they are having disruptive effects on previously established functional areas. The result of the shift to network-centric operations is a primary driver of convergence. This results in an upheaval of previously established functional areas. As was described above, computer network operations are complicating the management of EW, and this may also be the case for other functional areas such as PSYOPs. In addition to the other implications discussed below, specialists in many functional areas are required to adjust their thinking, and many may be dissatisfied with or resistant to some of the implications of convergence.

Legal and ethical considerations

The availability of information due to networked communication has increased drastically, and the open nature of social media has hastened this. A legal issue with conducting PYSOP on social media is that it may be difficult to restrict the operation to a specific target population due to its global availability. Therefore, it is highly likely that unintended audiences could access the content. This becomes problematic when there are legal restrictions on the population that is to be targeted. In addition, whilst a specific website may be for a specific nationality, it may be hosted in a different country that is off-limits. Therefore, targeting the audience of the webpage may fall into a grey area, legally speaking. The implications of this are that, when conducting cyber-enabled PSYOP, very careful legal analyses should be conducted to ensure that some legal boundary is not inadvertently crossed, which could result in severe embarrassment should a public outcry result.

The use of social media for mass influence or mass surveillance has additional ethical and legal concerns, particularly surrounding the misuse of these capabilities. An example is the use of social media by Pakistanis to threaten Indian citizens living in a particular area (Abbas 2012). These citizens then left the area in panic, causing a humanitarian crisis. Social media can be used to frame or misrepresent people as fake profiles can be created under names or legitimate profiles can be compromised with the specific goal of using these profiles to conduct misbehaviour for which the real person will then be blamed. Mass surveillance, legitimate or otherwise, has the real possibility of infringing on personal rights, particularly on privacy. The revelations that the US intelligence agencies were accessing a variety of communications resulted in a massive international outcry (Leyden 2013; Vijayan 2013). The very revelations of the communication intercepts and the information released by Bradley Manning and Wikileaks (Gragido & Pirc2011:192–195) fall within an ethical grey area. Their actions can be seen as a severe breach of security and irresponsible behaviour. However, their claimed motives and the views of their supporters are that this information is of public interest and illustrate governments' misuse of power. The releases were a form of IO in that the perpetrators were aiming to alter perceptions against the various governments and agencies. However, there seems to be an attitude that mass surveillance and cyber-espionage are allowable as long as you do not get caught: The US often complained about apparent Chinese incursion into their networks, and then the intercept revelations were made (Leyden 2013). French authorities are also showing anger. However, there are reports that they are involved in similar activities (Crowley 2013:12; Leyden 2013; Vijayan 2013). Many nations conduct espionage on competitors and allies alike (Crowley 2013:12), and the vast array of interconnected digital and telecommunications networks makes it easier to intercept or steal that information.

Legal jurisdiction over data and signals may become problematic. Data may cross multiple national boundaries between the sender and receiver, and with cloud computing, data may permanently reside outside of the owner's national borders. Likewise, radiated electromagnetic signals may cross national borders. These signals or data transfers may then be subjected to a wide variety of data or communication laws. If psychological or data attacks transit via a third nation, can this nation consider itself as under attack (when the attack is not aimed at it)? If this question can be answered, it still needs to be proved who the actual perpetrator is, as one of the problematic aspects of cyber-warfare is the difficulty of attribution (Liff 2012:412). This indicates that other attacks delivered through cyber-warfare tactics will probably be difficult to attribute. This perceived difficulty in attributing attacks may provide a feeling of invisibility and encourage activity which borders or crosses the line into being unethical whilst the perpetrators make the assumption that they will not be discovered.

Another issue with the difficulty of attribution is retaliation against an innocent party. This becomes more likely with the attacker actively framing a third party, hoping that the latter will get the blame. Should the retaliation against an attack be destructive in nature, it is imperative that the attribution information is accurate as this may inadvertently target innocent civilians whose system had been compromised by the attacker.

Organisational restructuring

In a business environment, Dennis and Durcikova (2012:109, 452) indicate that technological convergence of voice, video and data results in the audiovisual, IT and telecommunication departments of organisations merging to cope with the integrated technology. In a similar way, military units for PSYOP, EW, communication and cyber-warfare may find themselves strongly supporting each other, and joint-functionality IO cells may be formed for operations.

In a corporate setting, the corporate-communication and public-relations functions need to be concerned with the cyber-dimension in a more dynamic fashion. Not only do they need to run email communications to stakeholders and maintain traditional webpages, but they need to protect the organisation's image on social media. This entails an element of business intelligence where corporate-communication departments need to monitor the social media profiles run by them for negative comments. They also need to have intelligence on other potentially ‘hostile’ websites and profiles where disgruntled clients or employees could complain or behave in a manner damaging to the corporate image.

From an operational perspective, there is now a broader range of technology that need to be managed, integrated into operations and secured. Due to overlaps in domains and technology, the impact of operations may be broader than expected, and it may become difficult to contain operational effects. This then calls for improved business and competitive intelligence, particularly in the areas of assessing operational impacts. New committees may be required, or existing committees may need additional members with more or different specialities.

IO units or cells could therefore be comprised of a number of specialists working together, each of whom are experts in a specific area. Each specialist provides in-depth knowledge of their individual area. However, their expertise in other areas may be limited, and different views may cause conflict. An alternative is that the IO unit or cell is comprised of generalists who have grounding in multiple (or all) areas but are not necessarily experts in any. The generalists may be able to work together better due to common training and expertise. However, they may not have the depth of understanding that specialists would, potentially resulting in mistakes.

Porche *et al.* (2013) suggest separating the technical and psychological components in IO, and making commanders responsible for the integration of these when required. These authors indicate that the convergence of electronic warfare and cyber-operations requires personnel to have new specialities, pointing to a tendency to having generalists in technical areas (Porche *et al.* 2013:25). This will then effectively require two sets of specialists: those who deal with the technical aspects (cyber and EW) and those who deal with the psychological aspects (PSYOPs, strategic communications and public affairs). As the commanders are expected to be responsible for integration, they could then be considered as generalists.

Management techniques

The convergence of IO functional areas provides those with hostile intent with a variety of 'payloads' and delivery methods. At a strategic level, this may be managed to a certain degree by international policies and agreements. Operations including cyber and psychological aspects should be clarified in international law, particularly as it relates to armed conflict.

At the operational and tactical levels, the use of management information systems for command and control of cyber-operations and EW, such as the proposed systems described by Keller (2013) and the integrated cyber and electronic warfare system (Hatamoto 2013), will increase the efficiency of these operations by providing commanders with decision-support tools. These support systems can be utilised for both technical (EW and cyber) and psychologically based (including strategic communication) operations. These systems will necessitate efficient knowledge management and intelligence, which can be supported by the operations themselves (cyber-espionage and electronic intelligence) for data acquisition.

Training will need to be adjusted to cater for the converging operations. Those with technical skills will need to learn both EW and cyber-warfare concepts, as well as some of the psychological aspects of IO. The psychological operators will need to have combined training for PSYOP and more common public affairs practices with an introduction to the technical aspects, particularly the limitations regarding their use. The extent to which the cross-training is implemented will be determined by the structure of IO units and the extent to which they comprise of specialists or generalists as discussed above. Commanders, who will need to co-ordinate and ensure the integration of operations, will need to have extensive cross-training.

To ensure that operations do not cross any legal or ethical boundaries, legal advisors should be attached to IO cells or should advise commanders overseeing the activities.

CONCLUSION

IO consists of various functional areas, and a number of operational relationships exist amongst them. The overlaps are becoming larger due to the rapid uptake of wireless technology and social media. The article presented a qualitative document analysis to determine the nature of the convergence and the views thereof. There are many views on the convergence of functional areas in IO. Some strongly advocate that the cyber and electromagnetic domains should be kept separate. This article motivates that the domains can remain separate, but the operations and management thereof must be converging. This has management implications such as legal and ethical considerations for operations, organisational structure and training.

COMPETING INTERESTS

The authors declare that they have no financial or personal relationships which may have inappropriately influenced them in writing this article.

REFERENCES

1. Abbas, M., 2012, 'Web 2.0: Pakistan's new weapon?', CIOL.com, 21 August, viewed 23 August 2012, fromhttp://www.ciol.com/ciol/news/108402/web-pakistans-weapon

2. Alfreds, D., 2013, 'ANC admits website hack attack', News24, 14 June, viewed 17 June 2013, fromhttp://www.news24.com/Technology/News/ANC-admits-website-hack-attack-20130614

3. Amoroso, E.G., 2013, *Cyber attacks: Protecting national infrastructure*, Elsevier, Waltham, M.A.

4. Armistead, L., 2010, *Information operations matters*, Potomac Books, Washington, D.C.

5. Borque, J., 2008, 'A (pragmatic) future for joint electronic warfare: Does EW + CNO = Cyber?', *Journal of Electronic Defence* 31(9), 30–40.

6. Brazzoli, M.S., 2007, 'Future prospects of information warfare and particularly psychological operations', in L. le Roux (ed.), *South African army vision 2020*, pp. 217–232, Institute for Security Studies, Pretoria.

7. Chabrow, E., 2012, 'Aligning electronic and cyber warfare', viewed 11 July 2012, fromhttp://www.govinfosecurity.com/aligning-electronic-cyber-warfare-a-4930

8. Clifford, J., 2011, 'What electronic warriors should know about physics, language and concepts', *Journal of Electronic Defense* 34(3), 40–47.

9. Cloud, D.W., 2007, 'Integrated cyber defenses: Towards cyber defense doctrine', Master's dissertation, Naval Postgraduate School.

10. Cronin, B. & Crawford, H., 1999, 'Information warfare: Its application in military and civilian contexts', *The Information Society* 15(4), 257–263.

11. Crowley, M., 2013, 'Spy vs. spy', *Time Magazine*, 11 November, p. 12.

12. Daily News Correspondent, 2013, 'Hackers target Zimbabwe government', *Independent Online*, 14 June, viewed 14 June 2013, from http://www.iol.co.za/dailynews/news/hackers-target-zimbabwe-government-1.1532349#.UbtKI1EaKM8

13. Denning, D.E., 1999, *Information warfare and security*, Addison-Wesely, Boston.

14. Dennis, A. & Durcikova, A., 2012, *Fundamentals of business data communications*, 11th edn., Wiley, New York.

15. Department of the Army, 2014, *Field Manual 3–38: Cyber electromagnetic activities*, US Department of Defence, Washington, D.C.

16. Freedberg, S., 2013a, 'Army electronic warfare goes on the offensive: New tech awaits approval', 29 January, viewed 01 April 2013, from http://defense.aol.com/2013/01/29/army-electronic-warfare-new-tech/

17. Freedberg, S., 2013b, 'Gen. Hoss Cartwright talks immigration, cyber, China & Afghans with iPhones', 25 March, viewed 27 March 2013, from http://defense.aol.com/2013/03/25/gen-hoss-cartwright-talks-immigration-cyber-china-and-afghans-w/

18. Gould, J., 2013, 'New center, school to bring signals, cyber, EW together', *Army Times*, 25 June, viewed 07 September 2013, from http://www.armytimes.com/article/20130625/CAREERS/306250002/New-center-school-bring-signals-cyber-EW-together

19. Gragido, W. & Pirc, J., 2011, *Cybercrime and espionage*, Elsevier, Burlington.

20. Greenwald, G. & MacAskill, E., 2013, 'NSA Prism program taps in to user data of Apple, Google and others',*The Guardian*, 7 June, viewed 09 June 2013, from http://www.guardian.co.uk/world/2013/jun/06/us-tech-giants-nsa-data

21. Hahn, R., 2010, 'Physics of the cyber-EMS problem: Why we have the language wrong', *Journal of Electronic Defence* 33(11), 44–46.

22. Hatamoto, M., 2013, 'U.S. Army developing cyber, electronic war arsenal', *Daily Tech*, 31 October, viewed 13 November 2012, fromhttp://www.dailytech.com/US+Army+Developing+Cyber+Electronic+War+Arsenal/article33660.htm

23. Higgins, K.J., 2013, '"Red October" attacks: The new face of cyberespionage', *Dark Reading*, viewed 17 January 2013, from http://www.darkreading.com/attacks-breaches/red-october-attacks-the-new-face-of-cybe/240146237

24. Information Warfare Monitor, 2009, 'Tracking GhostNet: Investigating a Cyber Espionage Network', viewed 01 September 2009, from http://128.100.171.10/modules.php?op=modload&name=News&file=article&sid=2386

25. Joint Chiefs of Staff, 1998, *Joint publication 3–13: Information operations*, 09 October, US Department of Defence, Washington, D.C.

26. Kamzi, A., 2011, 'How anonymous emerged to occupy Wall Street', *The Guardian*, 27 September, viewed 06 October 2013, from http://www.guardian.co.uk/commentisfree/cifamerica/2011/sep/27/occupy-wall-street-anonymous

27. Keller, J., 2013, 'Industry: Get ready for spectrum warfare program to cover EW, optics, GPS, and cyber operations', *Military and Aerospace Electronics*, viewed 18 July 2013, fromhttp://www.militaryaerospace.com/articles/2013/07/usaf-answer-presolicitation.html

28. Knowles, J., 2013, 'Why two domains are better than one', *Journal of Electronic Defence* 36(5), 48–50.

29. Leyden, J., 2013, 'A post-Snowden US had better not squeal about Chinese cyber-spying', *The Register*, viewed 02 November 2013, fromhttp://www.theregister.co.uk/2013/11/01/snowden_effect_us_china_cyberespionage/

30. Liff, A.P., 2012, 'Cyberwar: A new "absolute weapon"?: The proliferation of cyberwarfare capabilities and interstate war', *Journal of Strategic Studies* 35(3), 401–428.

31. Maasdorp, F. & Du Plessis, W., 2012, 'Using a layered model to place EW in context within the informationsphere', *Proceedings of the 4th Workshop on ICT Uses in Warfare and the Safeguarding of Peace 2012 (IWSP 2012)*, 16 August 2012, Sandton, pp. 29–33.

32. Madrigal, A., 2011, 'The inside story of how Facebook responded to Tunisian hacks', *The Atlantic*, viewed 06 October 2013, from http://www.theatlantic.com/technology/archive/2011/01/the-inside-story-of-how-facebook-responded-to-tunisian-hacks/70044/#

33. Porche, I.R., Paul, C., York, M., Serena, C.C., Sollinger, J.M., Axelband, E. et al. 2013, *Redefining information warfare boundaries for an army in a wireless world*, RAND Institute, Santa Monica.

34. Prince, B., 2012, 'Project aims to detect cyber attacks using radio frequency', *Security Week*, viewed 21 November 2012, from http://www.securityweek.com/project-aims-detect-cyber-attacks-using-radio-frequency

35. Rouse, A. & Dick, M., 1994, 'The use of NUDIST, a computerized analytical tool, to support qualitative information systems research', *Information Technology* & *People* 7(3), 50–62.

36. Smith, R. & Knight, S., 2005, 'Applying electronic warfare solutions to network security', *Canadian Military Journal* 6(3), 49–58.

37. StrategyPage.com., 2009, 'Gaza cell phones targeted', 02 January, viewed 27 July 2009, fromhttp://www.strategypage.com/htmw/htiw/articles/20090102.aspx

38. Tubbs, B., 2013, 'SAPS hack spells negligence', viewed 22 May, from http://www.itweb.co.za/index.php?option=com_content&view=article&id=64268:SAPS-hack-spells-negligence&catid=265

39. Van Niekerk, B., 2011, 'Vulnerability analysis of modern ICT infrastructure from an information warfare perspective', PhD thesis, University of KwaZulu-Natal.

40. Van Niekerk, B., 2012, 'Tools for conducting operations on social media', South African National Defence Force Psychological Operations Expertise Register Workshop, 12 October, Pretoria.

41. Van Niekerk, B. & Maharaj, M.S., 2009, 'The future roles of electronic warfare in the information warfare spectrum', *Journal of Information Warfare* 8(3), 1–13.

42. Van Niekerk, B. & Maharaj, M.S., 2011, 'The information warfare life cycle model', *South African Journal of Information Management*13(1), 1–9.

43. Van Niekerk, B. & Maharaj, M.S., 2012, 'Mobile devices and the military: Useful tool or significant threat?',*Journal of Information Warfare* 11(2), 1–11.

44. Vijayan, J., 2013, 'Is French outrage against U.S. spying misplaced?', Computer World, viewed 25 October 2013, fromhttp://www.computerworld.com/s/article/9243414/Is_French_outrage_against_U.S._spying_misplaced

Chapter 12

Factors Influencing E-Collaboration for Knowledge Development and Innovation

Tendani J. Lavhengwa[1],Jacobus S. van der Walt[1],Eve M. Lavhengwa[2]

[1]*Department of Informatics, Tshwane University of Technology, South Africa*

[2]*Audit Services, Department of Minerals Resources, South Africa*

ABSTRACT

Background: Knowledge development and innovation are at the heart of the progress of academic and research institutions (ARIs) through individual and coordinated research projects. Collaboration initiatives remain a challenge for many researchers for a myriad of reasons which are further intensified by the many technology options that are available both freely and at varying prices. Although multiple theories were considered, the focus on electronic communication supported by the interest in how innovation is diffused and the richness of media motivated the focus on *diffusion of innovations* (DOI) and media richness theory (MRT).

Objectives: The objective was to develop a multi-dimensional matrix of e-collaboration factors for research institutions. This study investigated collaboration by ARIs while focusing on the supporting and enabling technologies.

Method: The grounded theory method (GTM) was adopted. E-collaboration literature was reviewed followed by data collection using observations, interviews and a blog. DOI and MRT were considered as theories that assist in the implementation of collaboration. A blog was developed as an e-collaboration platform to examine the emergent ideas and to collect data. Data was analysed through the coding method which led to the development of the multi-dimensional e-collaboration factors matrix.

Results: The findings reveal that e-collaboration has multiple factors that must be considered. Collaboration by participants was improved through knowledge development and innovation.

Conclusion: The multi-dimensional matrix of e-collaboration factors presented collaborators with a checklist that will enhance and improve their work. ARIs continue to collaborate at multiple levels depending on their needs and objectives.

INTRODUCTION

Collaboration, the development of knowledge and the improvement thereof for the sake of innovation are important for the research community. Because knowledge development is at the centre of research, academic and research institutions (ARIs) need to remain up to date with the changing electronic environment and requirements. Collaboration can entail one individual or one business, or it could entail multiple academics, institutions, nations or multinationals.

Soliman, Brown and Simoff (2005:372) assert that having a clearer understanding of what collaboration entails can help in with decision-making regarding future technology. The understanding of Soliman *et al.* (2005) of collaboration needs to be examined and relevant technologies must be identified for present and future use. This understanding will be improved by introducing a multi-dimensional matrix of e-collaboration factors.

Bettoni *et al.* (2011) observed that participative work is assumed to form part of academic activities. At times, this desire and need to work together is not natural; effort must be made and initiatives must be taken to make it possible. Anyangwe (2012) explain that higher-education institutions and further-education colleges must be prepared to share expertise. This statement is important since it joins ARIs to collaboration. When investigating the value of knowledge workers, Steyn and du Toit (2009:12) asserted that they were encouraged by participants' willingness to share their private knowledge stock and the willingness to using conversations as an approach to obtain the required knowledge. The motivation to focus on the use of electronic communication and blogs came from an investigation conducted by De Jager, Buitendag and Van der Walt (2012). De Jager *et al.*'s (2012) investigation was supported by the importance of discovering innovative collaborative knowledge. The need to share is thereby essential for various ARIs. There is also an inherent need to investigate supporting technology and derive improved levels of understanding.

De Jager *et al.* (2012) state that collaborative knowledge activities are necessary to generate vast quantities of knowledge within multiple domains. E-collaboration is considered as a contributor to knowledge development and innovation where similar challenges exist. This is more prevalent amongst ARIs where one of their core objectives is to conduct research. Understanding the factors that enable e-collaboration can assist in this challenge. The proposed matrix will also bridge the gap on the effective use of e-collaboration.

This article starts by discussing the literature related to driving forces, technology, e-collaboration concepts, knowledge development and ARIs. This is followed by examining the diffusion of innovation (DOI), media richness theory (MRT) and grounded theory method (GTM) as a grounding lens for analysis and discussion. The methodology adopted is GTM. The final outcome is to develop a multi-dimensional matrix of e-collaboration factors for research institutions.

The importance of driving forces and factors

Forces and factors are used interchangeably in this article. Driving forces or forces working towards achieving an objective are important. The focus is on driving forces for e-collaboration. The work of a number of authors work was reviewed to establish the driving forces as explained in the section below.

Caldwell (2009) listed five commonly used driving forces as follows: science and technology, economics, demographics, political and social. In support of driving forces, Bechina and Ndlela (2009) listed six factors that contribute to the effectiveness of knowledge management systems (KMS). These are leadership, training, clear business strategy, aligning business goal with technology, collaboration and adaptive culture.

Thomas (2013) asserts that collaboration works best for problems that have the following three characteristics measured as forces: There is no obvious solution, the problems lack structure, and the problems require collective volition.

These characteristics are significant since they also act as driving forces for individuals or groups to collaborate. It is also notable that the driving forces provide areas for further analysis.

Collaboration is, furthermore, listed as a driving force for KMS. The Canadian Health Services Research Foundation (CHSRF) (2003a:1) explains that knowledge brokering is a process of bringing people together and building relationships to make knowledge transfer more effective.

This linking of people can feature researchers, decision makers, practitioners and policy makers. The ability to bring people together and facilitate their interaction is listed as one of the basic skills required by a knowledge broker (CHSRF 2003a:1). The CHSRF (2003b:6) further asserts that there can be no standard definition of knowledge brokering since the job differs from context to context and is seen as an unfolding journey. The need to manage knowledge is central to improving the collaboration, hence the introduction of knowledge brokers who bring together knowledge participants. Technology is another primary driving force which connects collaborators from widely differing geographical areas and which presents multiple opportunities to collaborate.

Technology in the world today

Tools and technology have become central in facilitating interaction and communication. Tools and technology are used as interchangeable concepts. The days of seeing technology as a 'nice to have' have long passed. We need to acknowledge that technology has become pervasive. It is therefore essential for organisations and ARIs to be able to evaluate technology. Table 1 below shows a list of technological tools that can be used for collaboration, grouped by authors.

It is notable from Table 1 below that there is technology that is commonly used for collaboration. Knowing these can make it easier to work together with other participants to improve collaboration on projects.

In a detailed review, Sahin (2006) confirms the relationship between computer knowledge and the adoption of innovation. This relationship is why it is important to select the best technology for collaboration. It is also noted that the present level of computer knowledge as well as elements for e-collaboration require further investigation.

TABLE 1: Technology used for collaboration.

Author(s) name	List of technological tools
Blau (2011)	Online discussion groups, integrated calendaring and collaborative authoring tools.
Gartner (2010)	Wikis, blogs, instant messaging, collaborative office and crowdsourcing.
Hudson (2011:3)	Group-decision support systems, audio-conferencing and Internet-based web conferencing.
Vignarajah (2010)	E-mail, blog, micro-blogging, MySpace, Facebook, Twitter, LinkedIn, Podcast, Wiki, YouTube, text-messaging, web conferencing, iPhone and iPad.
Hill (2005)	Blogs, chat, email, list serve, message boards, online conferencing and threaded discussion.
Crow (2002)	E-mail exchange, drawing viewing sites (intranet and web-based), workflow and groupware software, teleconferencing and videoconferencing, web-hosted meetings and computer-aided design (CAD).

E-collaboration and key elementsKnowledge development related to innovation

Today's communication features e-collaboration as an integral component since work is mostly shared over electronic media, more specifically the Internet. With this in mind, it is important to develop an understanding of the origin and key elements of e-collaboration. Kock (2005) describe e-collaboration as the interaction of individuals engaged in a common task using electronic technology. Hudson (2011:3) presents a definition that encapsulates the objectives by denoting that collaboration occurs at any time where there are two or more people sharing complex information and knowledge building over the Internet.

Twinomurinzi (2007) explains that e-collaboration is the exchange of information, with the stakeholders playing a role in the outcome of the collaborative process. A more recent definition by Bettoni *et al.* (2011) described e-collaboration as a web-based group process of working, learning and sharing knowledge over distance in space and time.

Mindbuilt Technologies (2012) lists nine benefits of e-collaboration: increased efficiency, reduction in complexity, enhanced organisational intelligence, the development of stronger relationships, a reduction in travel costs, a reduction in long-distance phone calls, ease of installation and management, low overheads and a boost in employee morale.

The definitions and benefits listed above guided this article toward the proposed multi-dimensional matrix of e-collaboration factors.

Knowledge development related to innovation

E-collaboration is viewed as an improved mode of communication while using technology which varies from one interaction to another. Knowledge development and innovation can subsequently be listed as one of the outcomes of collaboration. The concepts contained in the preceding statement are discussed further in this section.

According to Rogers (2003), knowledge is the first step in the innovation-decision process. The other steps are persuasion, decision, implementation and confirmation. This endorses the relationship between knowledge and innovation. It also supports the idea that knowledge must precede innovation or be part thereof. Rogers (2003) further discusses the knowledge stage of the innovation-decision process which features three types of knowledge, namely awareness, how-to and principles. It is important to note that the 'how-to' knowledge can be associated with the selection of technology for use in an e-collaboration initiative.

Rogers (2003:12) describe innovation as an idea, practice or project that is perceived as new by an individual or other unit of adoption. This perception can be shared by one or many people and by one or many organisations or institutions. The definition of innovation is extensive and can also include products and processes for future implementation.

From the above discussion, it follows that knowledge development and innovation can be linked to the improvement of e-collaboration. E-collaboration technology therefore assists in the development of knowledge and innovation. Because the focus is on ARIs, it is imperative to discuss their connection to collaboration.

Academic and research institutions as a platform for collaboration

The Academy of Science of South Africa (ASSAf) (2010) classified universities into three categories: university, comprehensive university and university of technology. Statutory research bodies and institutions that fund research are also included since they are involved in research and funding. A number of collaboration initiatives were noted between universities and other research institutions and amongst universities themselves:

1. A memorandum of understanding was signed between UNISA and Mogale City (UNISA web 2013).

2. Planning, Research and Management issues relating to Mogale City Parks Service/Business excellence which may include operational and scientific management of resources, internal and external customer needs/satisfaction, management of biodiversity areas. (n.p.)

3. Mintek collaborates with the faculty of science at UJ (UJ web 2013): 'Collaboration between Mintek which is an organisation that works with the government on mineral technologies'. This collaboration connected a university, government and Mintek.

4. TUT web (2013) states that there is collaboration between the university and the French South African Institute of Technology (F'SATI), supported by the statement: '... a national asset that contributes to the creation of knowledge and prosperity as well as the transfer of technology in the Southern African region'. UCT

web (2013) describes the following: '... is an important step towards building drug discovery and development capabilities in Africa – and educating the next generation of drug-discovery scientists in Africa'.

The above sections were intended to illustrate and motivate the focus on ARIs and their collaborative relationships. The foundational theories for this study are discussed as a lens for analysis and further investigation.

Exploring theories related to e-collaboration

Theory is a good foundation for discussions on knowledge development and innovation. Theories are important in academia since they present underpinning arguments and researched views for further development. The section to follow discusses DOI, MRT and GTM.

Diffusion of innovation (DOI)

With e-collaboration being a form of innovation, we are discussing DOI elements related to the work by Rogers (2003). Rogers (2003:4) emphasises that DOI is a social process that is simultaneously a technical matter. One needs to understand the social surroundings and context in order to achieve a successful diffusion of innovation. Rogers (2003:5) further asserts that, in order to launch a self-gathering diffusion process, one must work with the correct participants from the start. This highlights the impact of that perceptions concerning DOI can have and how it needs to be managed. In order to improve and understand knowledge development for the sake of innovation, one needs to identify and influence the correct participants with whom to start. Rogers (2003:5) regarded diffusion as both the planned and spontaneous spread of new ideas and later listed the four elements of DOI. Table 2 lists these elements together with practical translations related to this investigation.

Media richness theory (MRT)

Daft and Lengel (1984) argued that MRT originates from information processing, developed by organisational scientists. Communication between people is affected by the fitness of the media and the characteristics of the communication task. Daft and Lengel (1984) further

TABLE 2: Diffusion of innovation elements and practical translations.

Element name	Description by Rogers (2003)	Practical translations related to this investigation
Innovation	'An idea, practice or project that is perceived as new by an individual or other unit of adoption'.	Working together while making use of technology such as e-collaboration.
Communication channels	'A process in which participants create and share information with one another in order to reach a mutual understanding'.	The different technological tools in place for collaboration to take place.
Time	'The impact and dimension of time on the diffusion'.	Time needs must be considered closely when addressing the diffusion of knowledge and innovation.
Social system	'A set of interrelated units engaged in joint problem solving to accomplish a common goal'.	These are the academic and research institutions involved in research. This social system can also feature independent researchers.

specified that MRT advances the notion that the richness or leanness of communication is an objective property of the communication media. These authors conclude by defining MRT as the ability to facilitate shared understanding within a time interval.

Miles (2014) asserts that people who use a communication channel most fit for their task will be more effective than people who use a communication channel that does not fit. This supports the importance of selecting the best-suited channel of communication for a collaboration experience.

Regarding media richness, Daft, Lengel and Trevino (1987) provided four criteria:

1. Capacity for immediate feedback: The medium facilitates quick convergence on a common interpretation.

2. Capacity to transmit multiple cues: An array of cues, including physical presence, voice inflections, body gestures, words, numbers and graphic symbols, facilitate interpretation and meaning rather than simply the transfer of information or data.

3. Language variety: Numbers and formulas provide greater precision, but natural language conveys a broader set of concepts.

4. Capacity of the medium to have a personal focus: This refers to either the conveyance of emotions and feelings or the ability of the medium tailored to the specific needs and perspectives of the receiver.

The criteria above were useful for determining the levels of media richness and also contribute towards measuring e-collaboration. Kock (2012) investigated media-richness and noted that it presents a partial confirmation that group members perceive e-collaboration as a relatively unnatural medium.The findings show that there is a perceived increase in the quality of outcomes in a group if the quality of member contribution increases. There is a higher departmental heterogeneity enabled by the low disruptiveness inherent in the e-collaboration medium used that is applicable to an asynchronous communication medium. The above theoretical work by Kock (2012) supports the argument that media richness makes a significant contribution to the improvement of communication.

In support of MRT, Dennis and Valacich (1999:5) describe media synchronicity as the extent to which individuals work together on the same activity at the same time. Multiple dimensions are presented: media characteristics, task, functions and communication processes.

Grounded theory method (GTM)

According to an early definition by Glaser and Strauss (1967), GTM refers to the discovery of theory from data systematically obtained and analysed in social research. Trochim (2000) further suggests that the research starts with generative questions since it is a complex iterative process. Strauss and Corbin (1990) described GTM as an approach where the data collection, analysis and theory stand in a reciprocal relationship with each other.

Research methodology and design

GTM was the approach selected for this study as described in the section on theory. The authors noted that the findings in each stage influence subsequent stages. The methodology adopted is as follows:

- Literature and theoretical reviews were conducted on the key topics and subjects.
- The notes from the primary data collected in the main study were reviewed. The data were collected through observations and interviews with information-rich participants, guided by the ideas below:
    - »Purposive sampling was selected as suitable as described by Welman, Kruger and Mitchell (2010:63) as an approach which allows researchers to rely on their experience, ingenuity and/or previous research findings.
    - »Patton (1990) asserts that information-rich cases are those from which one can learn a great deal about issues of central importance to the purpose of the research. From this term, purposeful sampling is introduced.
    - »The observations were conducted by the researcher who was physically present at the selected ARIs.

1. A blog was developed, based on the preceding phases and on those elements identified as significant and requiring further investigation.

2. An analysis of findings from multiple sources was done, using open coding described by McCallin, Nathanial and Scott (2011) as an activity that can take place while data are being collected.

3. A multi-dimensional matrix for e-collaboration factors was developed.

4. Discussions on the matrix factors and other related phenomenon began.

The methodology also used a blog for data collection and for further exploring the ideas. Lavhengwa and Van der Walt (2011) motivated the use of websites by asserting their value in being a platform for e-collaboration. One of the fundamentals identified is that Lavhengwa and Van der Walt (2011) argue that websites must be dynamic or interactive to improve research. Nardi, Schiano, Gumbrecht and Swartz (2004) asserts that blogs are a promising tool for knowledge management. Hill (2005) further supports blogs as a tool that enables collaborative learning over the Internet. Blogs were therefore selected for their collaborative qualities and their ability to work towards knowledge management and innovation. A blog was developed. Data was collected over a period of two years, and multiple findings emerged in support of and a focus on knowledge development and innovation. Below is a list of selected posts as questions that were used on the blog:

1. Mobile platforms and collaboration: How easy are these to identify, and how dominant are they as contributors towards e-collaboration?

2. What are the tools and technology for e-collaboration?

3. What drives and influences e-collaboration?

Table 3 shows the categories of participants for the research.

TABLE 3: Categories of respondents.

Category name	Units
Non-blog participants: (observation, emails, telephones and personal interviews)	31
Blog participants – page views	837†
Blog participants – published comments	73†

†, denotes that this is an on-going data collection and the number may have increased since they were last retrieved.

The non-blog data were collected through observation, e-mail, telephone and personal interviews conducted by the researcher. Observations allowed the data collection to take place while other researchers and academics were participating in academic activities. The significant level of interest can be noted from Table 3 by focusing on the numbers and the quality of the contributions which will be detailed in the findings section.

Findings and discussion

This section addresses the findings, followed by a discussion of each of the findings.

Empirical findings

The discussions that follow are related to the data collected from the participants as indicated in Table 3:

- Findings from non-blog participants: These were findings collected from 31 participants using multiple data-collection methods:
  - Mutual interest was found to be a motivator for collaboration.
  - Collaboration for learner support was listed as important for ARIs.
  - Collaboration agreements were noted as essential between the participants.

- Geographical distances had previously limited and hindered opportunities for collaboration. This challenge has been addressed with the introduction of e-collaboration.
- Technology bridged the physical and logical gap between collaborators through the introduction of virtual environments.
- »The knowledge economy is driven by electronic communication, thereby motivating an improvement in e-collaboration.
- The following collaboration technology was identified as prevalent: blogs, collaborative authoring tools, e-mail, telephone, fax and the Internet.
- Findings from the blog: These findings represent a list of the data collected from the e-collaboration blog:
- Auditors share knowledge at conferences, both physically and online
- A number of auditors make use of templates from their companies for conducting audits. These can be in a word, excel or web-based solution with access control.
- Internal auditors collaborate through audit tools such as TeaMate. This tool enables them to work together towards a common outcome. Reports are later generated from the multiple contributions made by the audit team.
- Internal and external auditors collaborate at the start of an audit project. They later work together towards the end when they have to generate and distribute reports. This collaboration is done primarily through email. The review notes or comments or responses are shared amongst the auditors.
- Audit findings are shared since they are the main deliverables or output for auditors.
- Technology such as blogs is examples of e-collaboration.
- Discussions are quite helpful, and we need to invite more people involved in e-learning, e-collaboration and related research projects.
- Self-service kiosks are also used for e-collaboration.

- Social media and mobile devices make it easy to e-collaborate.
- E-collaboration facilitates the collaboration of researchers from disadvantaged parts of the world with affluent peers from well-resourced universities.
- Eagerness to use technology: This means that the introduction of new technology determines the possibility of advanced e-collaboration.
- What follows is a list of key discussion points from the findings. The Internet and blogs can be useful as a technology for knowledge development. The introduction of e-collaboration discussions motivated and improved knowledge-development opportunities. Innovation was the next achievement when collaborators started to discuss new ideas on how to solve their challenges. Creating platforms for working together improves knowledge development.

There were indications that geographical distances are no longer a limiting factor for collaboration when technology is introduced.

In the next section, the outcome of the multi-dimensional matric for e-collaboration factors is introduced with all the key elements.

A multi-dimensional matrix for e-collaboration factors

A matrix is denoted as a grid which presents arrangements containing elements that require focus. In describing media synchronicity, Dennis and Valacich (1999:5) present multiple dimensions which motivated the key elements of the matrix of e-collaboration factors. These are presented in Table 4 and discussed thereafter.

Generic driving factors: There are generic driving forces that must be addressed. The relevant ones for this investigation are people, economics, financial, political, leadership and training. These are factors that affect general initiatives. The other dimensions of the matrix for e-collaboration factors can also be featured in this section, but they are in separate sections with comprehensive details.

TABLE 4: Matrix of e-collaboration factors.

Dimensions	Key elements identifies
Generic driving factors	Economics, finance, leadership, people, political and training.
Environmental and virtual	No geographical limitation and online access available anywhere.
Knowledge development and innovation	Identify knowledge areas of interest, identify relevant knowledge brokers and innovation must be initiated and improved.
Theoretical dimensions	Communication channels, time, social system, type and clarity of media in use for collaborating. An iterative process must be followed.
Tools and technology	Email, blogs, teleconferencing, Internet, social-media networks and online discussion groups.

Environmental and virtual: This refers to the environment or setting where the e-collaboration takes place.

This collaboration can take many forms, from physical to logical and virtual work spaces. E-collaboration must not be geographically limited to a specific area.

According to Chetty and Mearns (2012), virtual environments present effective knowledge-management platforms through communities of practice. These virtual environments assist when the participants are in geographically dispersed areas.

Knowledge development and innovation: Knowledge development can be defined by the identified objectives and projects that are in focus at a particular time. Collaborators must identify knowledge areas of interest.

The relevant knowledge brokers must be identified and approached to participate and assist in the development of the initiative. Improvement in people's work and new ideas on how to gain advantage over their competitors can be featured in this section. Innovation will result as reviews and reflections are made concerning the knowledge that is being developed.

Theoretical dimensions

The sections below focus on each of the theoretical dimensions.

Diffusion of innovation: An understanding must be developed on how collaboration and the related technology are being adopted. Relevant updates must follow and ensure that this process is on-going. The collaborator must consider the four elements of DOI listed by Rogers (2003:5), being innovation, communication channels, time and the social system. The diffusion of e-collaboration must not be assumed but planned, monitored and updated as changes immerge over time.

Media richness: The richness of media presents both benefits and challenges. Rich e-collaboration media is good as it can lead to an improvement in the clarity of the interaction and messages. However, e-collaboration is considered a relatively unnatural medium of communication (Kock 2012). This problem needs to be addressed, and the related concerns must be understood by the participants. Media richness must be considered very carefully and will differ from case to case. Communication must be adjusted to meet the needs of the specific collaboration initiative.

Grounded theory method: There must be a reciprocal approach in developing an e-collaboration experience. The process must start with generative questions, as suggested by Trochim (2000), without any prescriptive outcomes.

Tools and technology

Tools and technology have an important role to play in all electronic interactions. The selection of these must be done with the involvement of all relevant participants.

Figure 1 represents a diagrammatic view of the multi-dimensional e-collaboration factors matrix.

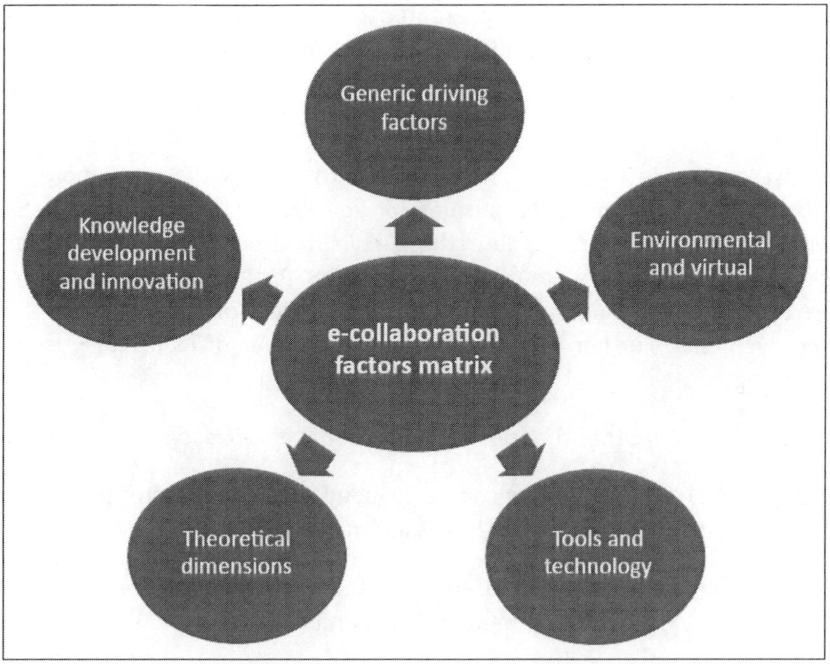

FIGURE 1: Multi-dimensional matrix of e-collaboration factors.

The abovementioned multi-dimensional matrix of e-collaboration factors does not prescribe an application sequence. Each individual project must set an order that is relevant for it. The users can decide what is of importance to them and start with that dimension.

The use and value of an e-collaboration matrix

This section initiates a discussion on the use of the matrix proposed. The primary purpose of the matrix is to assist in advancing a discussion on e-collaboration initiatives. The dimensions included are generic elements that would be adopted and addressed by collaborators. The matrix presents a high-level view of the focus of e-collaboration whilst focusing on research institutions. It is a guideline that can be applied, and it initiates a discussion for improved collaboration.

CONCLUSION AND PROSPECTS

Work on e-collaboration was presented for knowledge development and innovation. The article concludes that there is more work that can be done towards understanding e-collaboration, knowledge development and innovation.

Whilst the article focuses on ARIs in order to identify the e-collaboration factors, other environments (business or government) can be investigated with varying results to update the e-collaboration matrix. The factors identified in the article are a starting point for further investigation. They are also generic enough as a useful guide to initiate other discussions and improvements. Further work can feature a number of ideas such as the following:

1. Consider other theories as a lens for further investigation

2. Knowledge brokering can be included as a dimension of the factors matrix with additional details also featured.

3. The matrix can be used in a team as a guide for improved knowledge development and innovation.

4. The intention of investigation is that collaboration teams must consider and address all the dimensions in the matrix.

5. The factors matrix can be extended by adding other dimensions specific to a geographical area or community for an improved level of specialised understanding.

The final conclusion is that collaboration should be reviewed and adapted as time progresses. The technology in use must also be revised with the changing needs whilst guided by the e-collaboration matrix.

COMPETING INTERESTS

The authors declare that they have no financial or personal relationship(s) which may have inappropriately influenced them in writing this article.

AUTHORS' CONTRIBUTIONS

T.J.L. (Tshwane University of Technology) initiated the paper; allocated and co-ordinated all the roles for all authors; maintained and consolidated all contributions into the manuscript from start to finish. J.S.W. (Tshwane University of Technology) advised and supervised on all aspects of quality. E.L. (Department of Minerals Resources) contributed in the start-up discussions, assisted in language and logical advice to enhance the flow of the manuscript. E.L. also contributed in examining and grouping the empirical findings. The author further assessed the matrix developed.

REFERENCES

1. Academy of Science of South Africa (ASSAF), 2010, 'The PhD study: An evidence- based study on how to meet the demands for high-level skills in an emerging economy', Academy of Science of South Africa, Pretoria, viewed 09 October 2010, from http://www.assaf.org.za/wp-content/uploads/2010/10/40696- Boldesign-PHD-small-optimised.pdf

2. Anyangwe, E., 2012, 'HE in FE: Top tips and resources to promote collaboration. Guardian Professional', viewed 29 February 2012, from http://www.guardian.co.uk/higher-education-network/blog/2012/feb/09/he-in-fe-colleges?INTCMP=ILCNETTXT3487

3. Bechina, A.A. & Ndlela, M.N., 2009, 'Success factors in implementing knowledge based systems', *Electronic Journal of Knowledge Management* 7(2), 211–218.

4. Bettoni, M., Bernhard, W., Eggs, C. & Schiller, G., 2011, 'Participative faculty devel opment with an online course in e-collaboration', 6th International Conference on e-Learning, Kelowna.

5. Blau, I., 2011, 'E-collaboration within, between, and without institutions: Towards better functioning of online groups through networks', International Journal of e-Collaboration, 7(4), 37–60.

6. Caldwell, R.L., 2009, 'A new framework for anticipating the future', viewed 30 June 2013, fromhttp://cals.arizona.edu/~caldwell/docs/decision-framework-1-25-09.pdf

7. Canadian Health Services Research Foundation (CHSRF), 2003a, 'Preliminary report: The practice of knowledge brokering in Canada's health system', viewed 15 July 2013, from http://www.cfhi-fcass.ca/migrated/pdf/event_reports/brokers_final_e.pdf

8. Canadian Health Services Research Foundation (CHSRF), 2003b, 'CHSRF 2003 knowledge brokering workshop report', CHSRF knowledge brokering workshop, 02 October 2003, Montreal, viewed 15 July 2013, fromhttp://www.chsrf.ca/migrated/pdf/event_reports/Montreal_Report_e.pdf

9. Chetty, L. & Mearns, M., 2012, 'Using communities of practice towards the next level of knowledge management maturity', *SA Journal of Information Management* 14(1), Art. #503, 9 pages.

10. Crow, K., 2002, 'Collaboration: DRM associates', viewed 14 December 2012, from http://www.npd-solutions.com/collaboration.html

11. Daft, R.L. & Lengel, R.H., 1984, 'Information richness: A new approach to managerial behaviour and organization design',*Research in Organizational Behavior* 6, 191–233.

12. Daft, R.L., Lengel, R.H. & Trevino, L.K., 1987, 'Message equivocality, media selection, and manager performance: Implications for information systems', *Management Information Systems Quarterly* 11, 355–368.

13. De Jager, L., Buitendag, A.A.K. & Van der Walt, J.S., 2012, 'Presenting a framework for knowledge management within a web-enabled living lab', *SA Journal of Information Management* 14(1), Art. #506, 13 pages.

14. Dennis, A.R. & Valacich, J.S., 1999, 'Rethinking media richness: Towards a theory of media synchronicity', Proceedings of the 32nd Hawaii International Conference on System Sciences, 5–9 January, Hawaii.

15. Gartner, 2010, 'Gartner identifies the top 10 strategic technologies for 2011', Analysts examine latest industry trends during gartner symposium, ITxpo, Orlando, 17-21 October, viewed 09 May 2014, fromhttp://www.gartner.com/it/page.jsp?id=1454221

16. Glaser, B.G. & Strauss, A.L., 1967, 'The discovery of grounded theory', Aldine Publications, New York.

17. Hill, C., 2005, 'Collaboration and e-learning', Interactive Technology in Education Conference 865, viewed 15 August 2014, from http://itec.sfsu.edu/wp/865wp/s05_865_hill_collaboration.pdf

18. Hudson, A., 2011, 'The eight essential ingredients of small group e-collaboration technology', viewed 01 May 2013, fromhttp://www.grouputer.com/papers/8_essential_ingredients.pdf

19. Kock, N., 2005, 'What is e-collaboration', *International Journal of E-collaboration* 1(1), 1–7.

20. Kock, N., 2012, 'Unexpected outcomes of lean e-collaboration', viewed 25 February 2012, fromwww.cits.tamiu.edu/kockencycecollab/SampleManuscript2.doc

21. Lavhengwa, T.J. & Van der Walt, J.S., 2011, 'Websites: A platform for researchers and academic institutions to e-collaborate', Paper read at the 10th *Annual Information & Knowledge Management Conference*, 22–23 June 2011, Sandton, South Africa. McCallin, A., Nathanial, A. & Scott, H., 2011, 'Grounded theory online supporting GT researchers: Getting started', viewed 29 September 2011, from

22. http://www.groundedtheoryonline.com/getting-started.

23. Miles, J.A., 2014, 'Media richness theory: Management and organization theory', viewed 12 May 2014, fromhttp://citeseerx.ist.psu.edu/viewdoc/download?doi=10.1.1.108.7118&rep=rep1&type=pdf

24. Mindbuilt Technologies, 2012, *Why collaborate* ?, viewed 03 June 2012, from http://www.mindbuilt.com/ecs/reasons.php

25. Nardi, B., Schiano, D.J., Gumbrecht, M., Swartz, L. 2004, 'Why we blog', *Communications of the Association for Computing Machinery* 47(12), 41–46.

26. Patton, M., 1990, *Qualitative evaluation and research methods*, Sage, Beverly Hills.

27. Rogers, E.M., 2003, *Diffusion of innovations*, 5th edn., Free Press, New York.

28. Sahin, I., 2006, 'Detailed review of Rogers' *diffusion of innovations* theory and educational technology: Related studies based on Rogers' theory, *The Turkish Online Journal of Educational Technology* 5(2), 14–23.

29. Soliman, R., Braun, R. & Simoff, S., 2005, 'The essential ingredients of collaboration', Proceedings of the 2005 International Symposium on Collaborative technologies and Systems, p. 366, Saint Louis.

30. Steyn, P. & Du Toit, A., 2009, 'Maximising the value of knowledge workers', *South African Journal of Information Management* 11(1).

31. Strauss, A. & Corbin, J., 1990, *Basics of qualitative research: Grounded theory procedures and techniques*, Sage, London.

32. Thomas, R.J., 2013, 'The three essential ingredients of great collaborations', *Harvard Business Review* , viewed 12 May 2013, from http://blogs.hbr.org/cs/2011/06/the_three_essential_ingredient.html

33. Trochim, W., 2000, 'The research methods knowledge base', 2nd edn., Atomic Dog Publishing, Cincinnati.

34. Tshwane University of Technology (TUT web), 2013, 'TUT signs new agreement with French technology partner', viewed 30 June 2013, from http://www.tut.ac.za/News/Pages/TUTsignsnewagreementwithFrenchtechnologypartner.aspx

35. Twinomurinzi, H., 2007, 'E-collaboration for development innovations in rural communities: A South African experience', Proceedings of the 9th International Conference on Social

Implications of Computers in Developing Countries, São Paulo, Brazil, 28–30 May.

36. University Of Cape Town (UCT web), 2013, 'Collaboration with pharmaceutical giant to foster drug development in Africa', UCT Monday Paper, 32(3), viewed 30 June 2013, fromhttp://www.uct.ac.za/downloads/news.uct.ac.za/monpaper/mp32_03.pdf

37. University of Johannesburg (UJ web), 2013, 'Mintek – faculty of science collaboration', viewed 30 June 2013, fromhttp://www.uj.ac.za/en/research/research information/strategic partnerships/national partnerships/pages/mintek-facultyofsciencecollaboration.aspx

38. University of South Africa (UNISA web), 2013, 'Collaborations', viewed 30 June 2013, fromhttp://www.unisa.ac.za/Default.asp?Cmd=ViewContent&ContentID=19426

39. Vignarajah, A., 2010, 'Current trends in computer technology', Proceedings of the Global Business and Technology Association (GBATA), Twelfth Annual International Conference, pp. 911–915, Kruger National Park Vicinity, South Africa, 5-9 July.

40. Welman, J.C., Kruger, S.J. & Mitchell, B., 2010, *Research methodology*, 3rd edn., Oxford University Press, Cape Town.

Chapter 13

Beyond The Hype: Big Data Concepts, Methods, and Analytics

Amir Gandomi*, Murtaza Haider

Ted Rogers School of Management, Ryerson University, Toronto, Ontario M5B 2K3, Canadaa

ABSTRACT

Size is the first, and at times, the only dimension that leaps out at the mention of big data. This paper attempts to offer a broader definition of big data that captures its other unique and defining characteristics. The rapid evolution and adoption of big data by industry has leapfrogged the discourse to popular outlets, forcing the academic press to catch up. Academic journals in numerous disciplines, which will benefit from a relevant discussion of big data, have yet to cover the topic. This paper presents a consolidated description of big data by integrating definitions from practitioners and academics. The paper's primary focus is on the analytic methods used for big data. A particular distinguishing feature of this paper is its focus on analytics related to unstructured data, which constitute 95% of big data. This paper highlights the need to develop appropriate and efficient analytical methods to leverage massive volumes of heterogeneous data in unstructured text, audio, and video formats. This paper also reinforces the need to devise new tools for predictive analytics for structured big data. The statistical methods in practice were devised to infer from sample data. The heterogeneity, noise, and the massive size of structured big data calls for developing computationally efficient algorithms that may avoid big data pitfalls, such as spurious correlation.

Keywords: Big data analytics; Big data definition; Unstructured data analytics; Predictive analytics

1. INTRODUCTION

This paper documents the basic concepts relating to big data. It attempts to consolidate the hitherto fragmented discourse on what constitutes big data, what metrics define the size and other characteristics of big data, and what tools and technologies exist to harness the potential of big data.

From corporate leaders to municipal planners and academics, big data are the subject of attention, and to some extent, fear. The sudden rise of big data has left many unprepared. In the past, new technological developments first appeared in technical and academic publications. The knowledge and synthesis later seeped into other avenues of knowledge mobilization, including books. The fast evolution of big data technologies and the ready acceptance of the concept by public and private sectors left little time for the discourse to develop and mature in the academic domain. Authors and practitioners leapfrogged to books and other electronic media for immediate and wide circulation of their work on big data. Thus, one finds several books on big data, including *Big Data for Dummies*, but not enough fundamental discourse in academic publications.

The leapfrogging of the discourse on big data to more popular outlets implies that a coherent understanding of the concept and its nomenclature is yet to develop. For instance, there is little consensus around the fundamental question of how big the data has to be to qualify as 'big data'. Thus, there exists the need to document in the academic press the evolution of big data concepts and technologies.

A key contribution of this paper is to bring forth the oft-neglected dimensions of big data. The popular discourse on big data, which is dominated and influenced by the marketing efforts of large software and hardware developers, focuses on predictive analytics and structured data. It ignores the largest component of big data, which is unstructured and is available as audio, images, video, and unstructured text. It is estimated that the analytics-ready structured data forms only a small subset of big data. The unstructured data, especially data in video format, is the largest component of big data that is only partially archived.

This paper is organized as follows. We begin the paper by defining big data. We highlight the fact that size is only one of several dimensions of big data. Other characteristics, such as the frequency with which data are generated, are equally important in defining big data. We then expand the discussion on various types of big data, namely text, audio, video, and social media. We apply the analytics lens to the discussion on big data. Hence, when we discuss data in video format, we focus on methods and tools to analyze data in video format.

Given that the discourse on big data is contextualized in predictive analytics frameworks, we discuss how analytics have captured the imaginations of business and government leaders and describe the state-of-practice of a rapidly evolving industry. We also highlight the perils of big data, such as spurious correlation, which have hitherto escaped serious inquiry. The discussion has remained focused on correlation, ignoring the more nuanced and involved discussion on causation. We conclude by highlighting the expected developments to realize in the near future in big data analytics.

2. DEFINING BIG DATA

While it is ubiquitous today, however, 'big data' as a concept is nascent and has uncertain origins. Diebold (2012) argues that the term "big data ... probably originated in lunch-table conversations at Silicon Graphics Inc. (SGI) in the mid-1990s, in which John Mashey figured prominently". Despite the references to the mid-nineties, Fig. 1 shows that the term became widespread as recently as in 2011. The current hype can be attributed to the promotional initiatives by IBM and other leading technology companies who invested in building the niche analytics market.

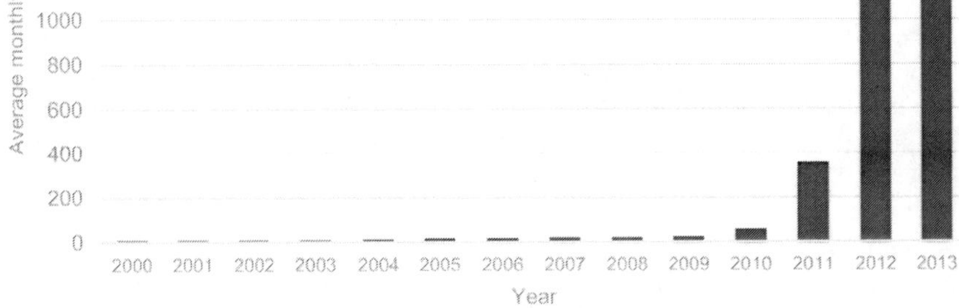

Figure 1. Frequency distribution of documents containing the term "big data" in ProQuest Research Library.

Big data definitions have evolved rapidly, which has raised some confusion. This is evident from an online survey of 154 C-suite global executives conducted by Harris Interactive on behalf of SAP in April 2012 ("Small and midsize companies look to make big gains with big data," 2012). Fig. 2 shows how executives differed in their understanding of big data, where some definitions focused on what it is, while others tried to answer what it does.

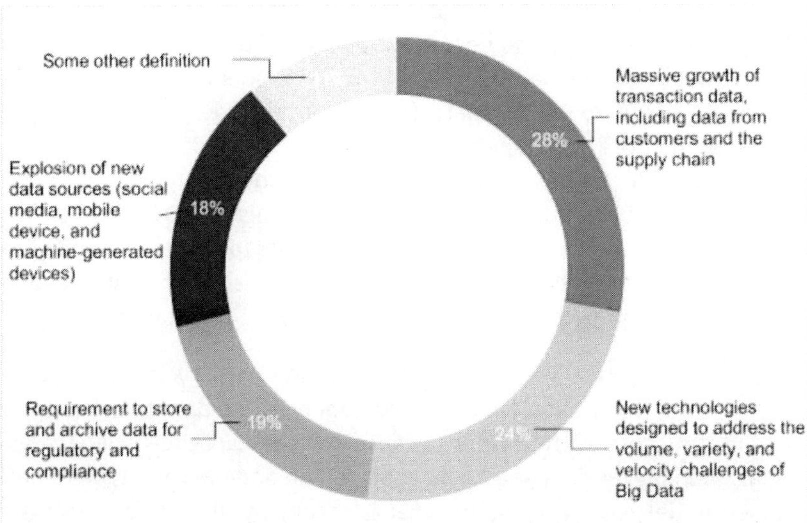

Figure 2. Definitions of big data based on an online survey of 154 global executives in April 2012.

Clearly, size is the first characteristic that comes to mind considering the question "what is big data?" However, other characteristics of big data have emerged recently. For instance, Laney (2001) suggested that *Volume*, *Variety*, and *Velocity* (or the *Three V's*) are the three dimensions of challenges in data management. The Three V's have emerged as a common framework to describe big data (Chen et al., 2012 and Kwon et al., 2014). For example, Gartner, Inc. defines big data in similar terms:

"*Big data is high-volume, high-velocity and high-variety information assets that demand cost-effective, innovative forms of information processing for enhanced insight and decision making*." (" Gartner IT Glossary, n.d.")

Similarly, TechAmerica Foundation defines big data as follows:

"Big data is a term that describes large volumes of high velocity, complex and variable data that require advanced techniques and technologies to enable the capture, storage, distribution, management, and analysis of the information." (TechAmerica Foundation's Federal Big Data Commission, 2012)

We describe the Three V's below.

Volume refers to the magnitude of data. Big data sizes are reported in multiple terabytes and petabytes. A survey conducted by IBM in mid-2012 revealed that just over half of the 1144 respondents considered datasets over one terabyte to be big data (Schroeck, Shockley, Smart, Romero-Morales, & Tufano, 2012). One terabyte stores as much data as would fit on 1500 CDs or 220 DVDs, enough to store around 16 million Facebook photographs. Beaver, Kumar, Li, Sobel, and Vajgel (2010) report that Facebook processes up to one million photographs per second. One petabyte equals 1024 terabytes. Earlier estimates suggest that Facebook stored 260 billion photos using storage space of over 20 petabytes.

Definitions of big data volumes are relative and vary by factors, such as time and the type of data. What may be deemed big data today may not meet the threshold in the future because storage capacities will increase, allowing even bigger data sets to be captured. In addition, the type of data, discussed under variety , defines what is meant by 'big'. Two datasets of the same size may require different data management technologies based on their type, e.g., tabular versus video data. Thus, definitions of big data also depend upon the industry. These considerations therefore make it impractical to define a specific threshold for big data volumes.

Variety refers to the structural heterogeneity in a dataset. Technological advances allow firms to use various types of structured, semi-structured, and unstructured data. Structured data, which constitutes only 5% of all existing data (Cukier, 2010), refers to the tabular data found in spreadsheets or relational databases. Text, images, audio, and video are examples of unstructured data, which sometimes lack the structural organization required by machines for analysis. Spanning a continuum between fully structured and unstructured data, the format of semi-structured data does not conform to strict standards. Extensible Markup Language (XML), a textual language for exchanging data on the Web, is a typical example of semi-structured data. XML documents contain user-defined data tags which make them machine-readable.

A high level of variety, a defining characteristic of big data, is not necessarily new. Organizations have been hoarding unstructured data from internal sources (e.g., sensor data) and external sources (e.g., social media). However, the emergence of new data management technologies and analytics, which enable organizations to leverage data in their business processes, is the innovative aspect. For instance, facial recognition technologies empower the brick-and-mortar retailers to acquire intelligence about store traffic, the age or gender composition of their customers, and their in-store movement patterns. This invaluable information is leveraged in decisions related to product promotions, placement, and staffing. Clickstream data provides a wealth of information about customer behavior and browsing patterns to online retailers. Clickstream advises on the timing and sequence of pages viewed by a customer. Using big data analytics, even small and medium-sized enterprises (SMEs) can mine massive volumes of semi-structured data to improve website designs and implement effective cross-selling and personalized product recommendation systems.

Velocity refers to the rate at which data are generated and the speed at which it should be analyzed and acted upon. The proliferation of digital devices such as smartphones and sensors has led to an unprecedented rate of data creation and is driving a growing need for real-time analytics and evidence-based planning. Even conventional retailers are generating high-frequency data. Wal-Mart, for instance, processes more than one million transactions per hour (Cukier, 2010). The data emanating from mobile devices and flowing through mobile apps produces torrents of information that can be used to generate real-time, personalized offers for everyday customers. This data provides sound information about customers, such as geospatial location, demographics, and past buying patterns, which can be analyzed in real time to create real customer value.

Given the soaring popularity of smartphones, retailers will soon have to deal with hundreds of thousands of streaming data sources that demand real-time analytics. Traditional data management systems are not capable of handling huge data feeds instantaneously. This is where big data technologies come into play. They enable firms to create real-time intelligence from high volumes of 'perishable' data.

In addition to the three V's, other dimensions of big data have also been mentioned. These include:

- *Veracity*. IBM coined *Veracity* as the fourth V, which represents the unreliability inherent in some sources of data. For example, customer sentiments in social media are uncertain in nature, since they entail human judgment. Yet they contain valuable information. Thus the need to deal with imprecise and uncertain data is another facet of big data, which is addressed using tools and analytics developed for management and mining of uncertain data.

- *Variability* (*and complexity*). SAS introduced *Variability* and *Complexity* as two additional dimensions of big data. Variability refers to the variation in the data flow rates. Often, big data velocity is not consistent and has periodic peaks and troughs. Complexity refers to the fact that big data are generated through a myriad of sources. This imposes a critical challenge: the need to connect, match, cleanse and transform data received from different sources.
- *Value*. Oracle introduced *Value* as a defining attribute of big data. Based on Oracle's definition, big data are often characterized by relatively "low value density". That is, the data received in the original form usually has a low value relative to its volume. However, a high value can be obtained by analyzing large volumes of such data.

The relativity of big data volumes discussed earlier applies to all dimensions. Thus, universal benchmarks do not exist for volume, variety, and velocity that define big data. The defining limits depend upon the size, sector, and location of the firm and these limits evolve over time. Also important is the fact that these dimensions are not independent of each other. As one dimension changes, the likelihood increases that another dimension will also change as a result. However, a 'three-V tipping point' exists for every firm beyond which traditional data management and analysis technologies become inadequate for deriving timely intelligence. The Three-V tipping point is the threshold beyond which firms start dealing with big data. The firms should then trade-off the future value expected from big data technologies against their implementation costs.

3. BIG DATA ANALYTICS

Big data are worthless in a vacuum. Its potential value is unlocked only when leveraged to drive decision making. To enable such evidence-based decision making, organizations need efficient processes to turn high volumes of fast-moving and diverse data into meaningful insights. The overall process of extracting insights from big data can be broken down into five stages (Labrinidis & Jagadish, 2012), shown in Fig. 3. These five stages form the two main sub-processes: data management and analytics. Data management involves processes and supporting technologies to acquire and store data and to prepare and retrieve it for analysis. Analytics, on the other hand, refers to techniques used to analyze and acquire intelligence from big data. Thus, big data analytics can be viewed as a sub-process in the overall process of 'insight extraction' from big data.

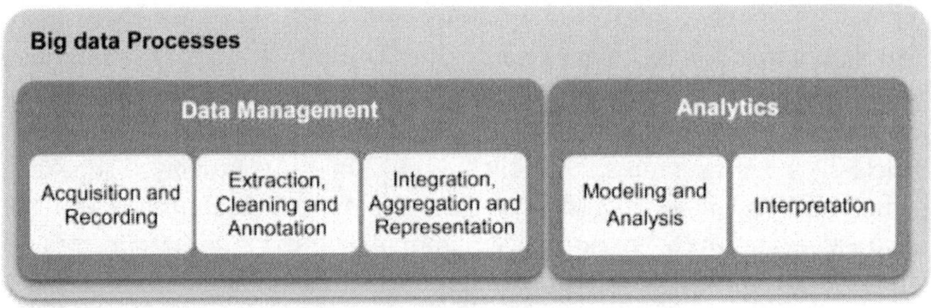

Figure 3. Processes for extracting insights from big data.

In the following sections, we briefly review big data analytical techniques for structured and unstructured data. Given the breadth of the techniques, an exhaustive list of techniques is beyond the scope of a single paper. Thus, the following techniques represent a relevant subset of the tools available for big data analytics.

3.1. Text analytics

Text analytics (text mining) refers to techniques that extract information from textual data. Social network feeds, emails, blogs, online forums, survey responses, corporate documents, news, and call center logs are examples of textual data held by organizations. Text analytics involve statistical analysis, computational linguistics, and machine learning. Text analytics enable businesses to convert large volumes of human generated text into meaningful summaries, which support evidence-based decision-making. For instance, text analytics can be used to predict stock market based on information extracted from financial news (Chung, 2014). We present a brief review of text analytics methods below.

Information extraction (*IE*) techniques extract structured data from unstructured text. For example, IE algorithms can extract structured information such as drug name, dosage, and frequency from medical prescriptions. Two sub-tasks in IE are Entity Recognition (ER) and Relation

Extraction (RE) (Jiang, 2012). ER finds names in text and classifies them into predefined categories such as person, date, location, and organization. RE finds and extracts semantic relationships between entities (e.g., persons, organizations, drugs, genes, etc.) in the text. For example, given the sentence "Steve Jobs co-founded Apple Inc. in 1976", an RE system can extract relations such as FounderOf [Steve Jobs, Apple Inc.] or FoundedIn [Apple Inc., 1976].

Text summarization techniques automatically produce a succinct summary of a single or multiple documents. The resulting summary conveys the key information in the original text(s). Applications include scientific and news articles, advertisements, emails, and blogs. Broadly speaking, summarization follows two approaches: the extractive approach and the abstractive approach. In extractive summarization, a summary is created from the original text units (usually sentences). The resulting summary is a subset of the original document. Based on the extractive approach, formulating a summary involves determining the salient units of a text and stringing them together. The importance of the text units is evaluated by analyzing their location and frequency in the text. Extractive summarization techniques do not require an 'understanding' of the text. In contrast, abstractive summarization techniques involve extracting semantic information from the text. The summaries contain text units that are not necessarily present in the original text. In order to parse the original text and generate the summary, abstractive summarization incorporates advanced Natural Language Processing (NLP) techniques. As a result, abstractive systems tend to generate more coherent summaries than the extractive systems do (Hahn & Mani, 2000). However, extractive systems are easier to adopt, especially for big data.

Question answering (*QA*) techniques provide answers to questions posed in natural language. Apple's Siri and IBM's Watson are examples of commercial QA systems. These systems have been implemented in healthcare, finance, marketing, and education. Similar to abstractive summarization, QA systems rely on complex NLP techniques. QA techniques are further classified into three categories: the information retrieval (IR)-based approach, the knowledge-based approach, and the hybrid approach. IR-based QA systems often have three sub-components. First is the *question processing*, used to determine details, such as the question type, question focus, and the answer type, which are used to create a query. Second is *document processing* which is used to retrieve relevant pre-written passages from a set of existing documents using the query formulated in question processing. Third is *answer processing*, used to extract candidate answers from the output of the previous component, rank them, and return the highest-ranked candidate as the output of the QA system. Knowledge-based QA systems generate a semantic description of the question, which is then used to query structured resources. The Knowledge-based QA systems are particularly useful for restricted domains, such as tourism, medicine, and transportation, where large volumes of pre-written documents do not exist. Such domains lack data redundancy, which is required for IR-based QA systems. Apple's Siri is an example of a QA system that exploits the knowledge-based approach. In hybrid QA systems, like IBM's Watson, while the question is semantically analyzed, candidate answers are generated using the IR methods.

Sentiment analysis (*opinion mining*) techniques analyze opinionated text, which contains people's opinions toward entities such as products, organizations, individuals, and events. Businesses are increasingly capturing more data about their customers' sentiments that has led to the proliferation of sentiment analysis (Liu, 2012). Marketing, finance, and the political and social sciences are the major application areas of sentiment analysis. Sentiment analysis techniques are further divided into three sub-groups, namely document-level, sentence-level, and aspect-based. Document-level techniques determine whether the whole document expresses a negative or a positive sentiment. The assumption is that the document contains sentiments about a single entity. While certain techniques categorize a document into two classes, negative and positive, others incorporate more sentiment classes (like the Amazon's five-star system) (Feldman, 2013). Sentence-level techniques attempt to determine the polarity of a single sentiment about a known entity expressed in a single sentence. Sentence-level techniques must first distinguish subjective sentences from objective ones. Hence, sentence-level techniques tend to be more complex compared to document-level techniques. Aspect-based techniques recognize all sentiments within a document and identify the aspects of the entity to which each sentiment refers. For instance, customer product reviews usually contain opinions about different aspects (or features) of a product. Using aspect-based techniques, the vendor can obtain valuable information about different features of the product that would be missed if the sentiment is only classified in terms of polarity.

3.2. Audio analytics

Audio analytics analyze and extract information from unstructured audio data. When applied to human spoken language, audio analytics is also referred to as *speech analytics*. Since these techniques have mostly been applied to spoken audio, the terms audio analytics and speech analytics are often used interchangeably. Currently, customer call centers and healthcare are the primary application areas of audio analytics.

Call centers use audio analytics for efficient analysis of thousands or even millions of hours of recorded calls. These techniques help improve customer experience, evaluate agent performance, enhance sales turnover rates, monitor compliance with different policies (e.g., privacy and security policies), gain insight into customer behavior, and identify product or service issues, among many other tasks. Audio analytics systems can be designed to analyze a live call, formulate cross/up-selling recommendations based on the customer's past and present interactions, and provide feedback to agents in real time. In addition, automated call centers use the Interactive Voice Response (IVR) platforms to identify and handle frustrated callers.

In healthcare, audio analytics support diagnosis and treatment of certain medical conditions that affect the patient's communication patterns (e.g., depression, schizophrenia, and cancer) (Hirschberg, Hjalmarsson, & Elhadad, 2010). Also, audio analytics can help analyze an infant's cries, which contain information about the infant's health and emotional status (Patil, 2010). The vast amount of data recorded through speech-driven clinical documentation systems is another driver for the adoption of audio analytics in healthcare.

Speech analytics follows two common technological approaches: the transcript-based approach (widely known as large-vocabulary continuous speech recognition, LVCSR) and the phonetic-based approach. These are explained below.

- *LVCSR systems* follow a two-phase process: indexing and searching. In the first phase, they attempt to transcribe the speech content of the audio. This is performed using automatic speech recognition (ASR) algorithms that match sounds to words. The words are identified based on a predefined dictionary. If the system fails to find the exact word in the dictionary, it returns the most similar one. The output of the system is a searchable index file that contains information about the sequence of the words spoken in the speech. In the second phase, standard text-based methods are used to find the search term in the index file.

- *Phonetic-based systems* work with sounds or *phonemes*. Phonemes are the perceptually distinct units of sound in a specified language that distinguish one word from another (e.g., the phonemes/k/and/b/differentiate the meanings of “cat” and “bat”). Phonetic-based systems also consist of two phases: phonetic indexing and searching. In the first phase, the system translates the input speech into a sequence of phonemes. This is in contrast to LVCSR systems where the speech is converted into a sequence of words. In the second phase, the system searches the output of the first phase for the phonetic representation of the search terms.

3.3. Video analytics

Video analytics, also known as video content analysis (VCA), involves a variety of techniques to monitor, analyze, and extract meaningful information from video streams. Although video analytics is still in its infancy compared to other types of data mining (Panigrahi, Abraham, & Das, 2010), various techniques have already been developed for processing real-time as well as pre-recorded videos. The increasing prevalence of closed-circuit television (CCTV) cameras and the booming popularity of video-sharing websites are the two leading contributors to the growth of computerized video analysis. A key challenge, however, is the sheer size of video data. To put this into perspective, one second of a high-definition video, in terms of size, is equivalent to over 2000 pages of text (Manyika et al., 2011). Now consider that 100 hours of video are uploaded to YouTube every minute (YouTube Statistics, n.d.).

Big data technologies turn this challenge into opportunity. Obviating the need for cost-intensive and risk-prone manual processing, big data technologies can be leveraged to automatically sift through and draw intelligence from thousands of hours of video. As a result, the big data technology is the third factor that has contributed to the development of video analytics.

The primary application of video analytics in recent years has been in automated security and surveillance systems. In addition to their high cost, labor-based surveillance systems tend to be less effective than automatic systems (e.g., Hakeem et al., 2012 report that security personnel cannot remain focused on surveillance tasks for more than 20 minutes). Video analytics can efficiently and effectively perform surveillance functions such as detecting breaches of restricted zones, identifying objects removed or left unattended, detecting loitering in a specific area, recognizing suspicious activities, and detecting camera tampering, to name a few. Upon detection of a threat, the surveillance system may notify security personnel in real time or trigger an automatic action (e.g., sound alarm, lock doors, or turn on lights).

The data generated by CCTV cameras in retail outlets can be extracted for business intelligence. Marketing and operations management are the primary application areas. For instance, smart algorithms can collect demographic information about customers, such as age, gender, and ethnicity. Similarly, retailers can count the number of customers, measure the time they stay in the store, detect their movement patterns, measure their dwell time in different areas, and monitor queues in real time. Valuable insights can be obtained by correlating this information with customer demographics to drive decisions for product placement, price, assortment optimization, promotion design, cross-selling, layout optimization, and staffing.

Another potential application of video analytics in retail lies in the study of buying behavior of groups. Among family members who shop together, only one interacts with the store at the cash register, causing the traditional systems to miss data on buying patterns of other members. Video analytics can help retailers address this missed opportunity by providing information about the size of the group, the group's demographics, and the individual members' buying behavior.

Automatic video indexing and retrieval constitutes another domain of video analytics applications. The widespread emergence of online and offline videos has highlighted the need to index multimedia content for easy search and retrieval. The indexing of a video can be performed based on different levels of information available in a video including the metadata, the soundtrack, the transcripts, and the visual content of the video. In the metadata-based approach, relational database management systems (RDBMS) are used for video search and retrieval. Audio analytics and text analytics techniques can be applied to index a video based on the associated soundtracks and transcripts, respectively. A comprehensive review of approaches and techniques for video indexing is presented in Hu, Xie, Li, Zeng, and Maybank (2011).

In terms of the system architecture, there exist two approaches to video analytics, namely server-based and edge-based:

- *Server-based architecture*. In this configuration, the video captured through each camera is routed back to a centralized and dedicated server that performs the video analytics. Due to bandwidth limits, the video generated by the source is usually compressed by reducing the frame rates and/or the image resolution. The resulting loss of information can affect the accuracy of the analysis. However, the server-based approach provides economies of scale and facilitates easier maintenance.

- *Edge-based architecture*. In this approach, analytics are applied at the 'edge' of the system. That is, the video analytics is performed locally and on the raw data captured by the camera. As a result, the entire content of the video stream is available for the analysis, enabling a more effective content analysis. Edge-based systems, however, are more costly to maintain and have a lower processing power compared to the server-based systems.

3.4. Social media analytics

Social media analytics refer to the analysis of structured and unstructured data from social media channels. Social media is a broad term encompassing a variety of online platforms that allow users to create and exchange content. Social media can be categorized into the following types: Social networks (e.g., Facebook and LinkedIn), blogs (e.g., Blogger and WordPress), microblogs (e.g., Twitter and Tumblr), social news (e.g., Digg and Reddit), social bookmarking (e.g., Delicious and StumbleUpon), media sharing (e.g., Instagram and YouTube), wikis (e.g., Wikipedia and Wikihow), question-and-answer sites (e.g., Yahoo! Answers and Ask.com) and review sites (e.g., Yelp, TripAdvisor) (Barbier and Liu, 2011 and Gundecha and Liu, 2012). Also, many mobile apps, such as Find My Friend, provide a platform for social interactions and, hence, serve as social media channels.

Although the research on social networks dates back to early 1920s, nevertheless, social media analytics is a nascent field that has emerged after the advent of Web 2.0 in the early 2000s. The key characteristic of the modern social media analytics is its data-centric nature. The research on social media analytics spans across several disciplines, including psychology, sociology, anthropology, computer science, mathematics, physics, and economics. Marketing has been the primary application of social media analytics in recent years. This can be attributed to the widespread and growing adoption of social media by consumers worldwide (He, Zha, & Li, 2013), to the extent that Forrester Research, Inc., projects social media to be the second-fastest growing marketing channel in the US between 2011 and 2016 (VanBoskirk, Overby, & Takvorian, 2011).

User-generated content (e.g., sentiments, images, videos, and bookmarks) and the relationships and interactions between the network entities (e.g., people, organizations, and products) are the two sources of information in social media. Based on this categorization, the social media analytics can be classified into two groups:

- *Content-based analytics*. Content-based analytics focuses on the data posted by users on social media platforms, such as customer feedback, product reviews, images, and videos. Such content on social media is often voluminous, unstructured, noisy, and dynamic. Text, audio, and video analytics, as discussed earlier, can be applied to derive insight from such data. Also, big data technologies can be adopted to address the data processing challenges.

- *Structure-based analytics*. Also referred to as *social network analytics*, this type of analytics are concerned with synthesizing the structural attributes of a social network and extracting intelligence from the relationships among the participating entities. The structure of a social network is modeled through a set of nodes and edges, representing participants and relationships, respectively. The model can be visualized as a graph composed of the nodes and the edges. We review two types of network graphs, namely *social graphs* and *activity graphs* (Heidemann, Klier, & Probst, 2012). In social graphs, an edge between a pair of nodes only signifies the existence of a link (e.g., friendship) between the corresponding entities. Such graphs can be mined to identify communities or determine hubs (i.e., the users with a relatively large number of direct and indirect social links). In activity networks, however, the edges represent actual interactions between any pair of nodes. The interactions involve exchanges of information (e.g., likes and comments). Activity graphs are preferable to social graphs, because an active relationship is more relevant to analysis than a mere connection.

Various techniques have recently emerged to extract information from the structure of social networks. We briefly discuss these below.

Community detection, also referred to as community discovery, extracts implicit communities within a network. For online social networks, a community refers to a sub-network of users who interact more extensively with each other than with the rest of the network. Often containing millions of nodes and edges, online social networks tend to be colossal in size. Community detection helps to summarize huge networks, which then facilitates uncovering existing behavioral patterns and predicting emergent properties of the network. In this regard, community detection is similar to clustering (Aggarwal, 2011), a data mining technique used to partition a data set into disjoint subsets based on the similarity of data points. Community detection has found several application areas, including marketing and the World Wide Web (Parthasarathy, Ruan, & Satuluri, 2011). For example, community detection enables firms to develop more effective product recommendation systems.

Social influence analysis refers to techniques that are concerned with modeling and evaluating the influence of actors and connections in a social network. Naturally, the behavior of an actor in a social network is affected by others. Thus, it is desirable to evaluate the participants' influence, quantify the strength of connections, and uncover the patterns of influence diffusion in a network. Social influence analysis techniques can be leveraged in viral marketing to efficiently enhance brand awareness and adoption.

A salient aspect of social influence analysis is to quantify the importance of the network nodes. Various measures have been developed for this purpose, including degree centrality, betweenness centrality, closeness centrality, and eigenvector centrality (for more details refer to Tang & Liu, 2010). Other measures evaluate the strength of connections represented by edges or model the spread of influence in social networks. The Linear Threshold Model (LTM) and Independent Cascade Model (ICM) are two well-known examples of such frameworks (Sun & Tang, 2011).

Link prediction specifically addresses the problem of predicting future linkages between the existing nodes in the underlying network. Typically, the structure of social networks is not static and continuously grows through the creation of new nodes and edges. Therefore, a natural goal is to understand and predict the dynamics of the network. Link prediction techniques predict the occurrence of interaction, collaboration, or influence among entities of a network in a specific time interval. Link prediction techniques outperform pure chance by factors of 40–50, suggesting that the current structure of the network surely contains latent information about future links (Liben-Nowell & Kleinberg, 2003).

In biology, link prediction techniques are used to discover links or associations in biological networks (e.g., protein–protein interaction networks), eliminating the need for expensive experiments (Hasan & Zaki, 2011). In security, link prediction helps to uncover potential collaborations in terrorist or criminal networks. In the context of online social media, the primary application of link prediction is in the development of recommendation systems, such as Facebook's "People You May Know", YouTube's "Recommended for You", and Netflix's and Amazon's recommender engines.

3.5. Predictive analytics

Predictive analytics comprise a variety of techniques that predict future outcomes based on historical and current data. In practice, predictive analytics can be applied to almost all disciplines – from predicting the failure of jet engines based on the stream of data from several thousand sensors, to predicting customers' next moves based on what they buy, when they buy, and even what they say on social media.

At its core, predictive analytics seek to uncover patterns and capture relationships in data. Predictive analytics techniques are subdivided into two groups. Some techniques, such as moving averages, attempt to discover the historical patterns in the outcome variable(s) and extrapolate them to the future. Others, such as linear regression, aim to capture the interdependencies between outcome variable(s) and explanatory variables, and exploit them to make predictions. Based on the underlying methodology, techniques can also be categorized into two groups: regression techniques (e.g., multinomial logit models) and machine learning techniques (e.g., neural networks). Another classification is based on the type of outcome variables: techniques such as linear regression address continuous outcome variables (e.g., sale price of houses), while others such as Random Forests are applied to discrete outcome variables (e.g., credit status).

Predictive analytics techniques are primarily based on statistical methods. Several factors call for developing new statistical methods for big data. First, conventional statistical methods are rooted in statistical significance: a small sample is obtained from the population and the result is compared with chance to examine the significance of a particular relationship. The conclusion is then generalized to the entire population. In contrast, big data samples are massive and represent the majority of, if not the entire, population. As a result, the notion of statistical significance is not that relevant to big data. Secondly, in terms of computational efficiency, many conventional methods for small samples do not scale up to big data. The third factor corresponds to the distinctive features inherent in big data: heterogeneity, noise accumulation, spurious correlations, and incidental endogeneity (Fan, Han, & Liu, 2014). We describe these below.

- *Heterogeneity*. Big data are often obtained from different sources and represent information from different sub-populations. As a result, big data are highly heterogeneous. The sub-population data in small samples are deemed outliers because of their insufficient frequency. However, the sheer size of big data sets creates the unique opportunity to model the heterogeneity arising from sub-population data, which would require sophisticated statistical techniques.

- *Noise accumulation*. Estimating predictive models for big data often involves the simultaneous estimation of several parameters. The accumulated estimation error (or noise) for different parameters could dominate the magnitudes of variables that have true effects within the model. In other words, some variables with significant explanatory power might be overlooked as a result of noise accumulation.

- *Spurious correlation*. For big data, spurious correlation refers to uncorrelated variables being falsely found to be correlated due to the massive size of the dataset.Fan and Lv (2008) show this phenomenon through a simulation example, where the correlation coefficient between independent random variables is shown to increase with the size of the dataset. As a result, some variables that are scientifically unrelated (due to their independence) are erroneously proven to be correlated as a result of high dimensionality.

- *Incidental endogeneity*. A common assumption in regression analysis is the exogeneity assumption: the explanatory variables, or predictors, are independent of the residual term. The validity of most statistical methods used in regression analysis depends on this assumption. In other words, the existence of incidental endogeneity (i.e., the dependence of the residual term on some of the predictors) undermines the validity of the statistical methods used for regression analysis. Although the exogeneity assumption is usually met in small samples, incidental endogeneity is commonly present in big data. It is worthwhile to mention that, in contrast to spurious correlation, incidental endogeneity refers to a genuine relationship between variables and the error term.

The irrelevance of statistical significance, the challenges of computational efficiency, and the unique characteristics of big data discussed above highlight the need to develop new statistical techniques to gain insights from predictive models.

4. CONCLUDING REMARKS

The objective of this paper is to describe, review, and reflect on big data. The paper first defined what is meant by big data to consolidate the divergent discourse on big data. We presented various definitions of big data, highlighting the fact that size is only one dimension of big data. Other dimensions, such as velocity and variety are equally important. The paper's primary focus has been on analytics to gain valid and valuable insights from big data. We highlight the point that predictive analytics, which deals mostly with structured data, overshadows other forms of analytics applied to unstructured data, which constitutes 95% of big data. We reviewed analytics techniques for text, audio, video, and social media data, as well as predictive analytics. The paper makes the case for new statistical techniques for big data to address the peculiarities that differentiate big data from smaller data sets. Most statistical methods in practice have been devised for smaller data sets comprising samples.

Technological advances in storage and computations have enabled cost-effective capture of the informational value of big data in a *timely* manner. Consequently, one observes a proliferation in real-world adoption of analytics that were not economically feasible for large-scale applications prior to the big data era. For example, sentiment analysis (opinion mining) have been known since the early 2000s (Pang & Lee, 2008). However, big data technologies enabled businesses to adopt sentiment analysis to glean useful insights from millions of opinions shared on social media. The processing of unstructured text fueled by the massive influx of social media data is generating business value by adopting conventional (pre-big data) sentiment analysis techniques, which may not be ideally suited to leverage big data.

Although major innovations in analytical techniques for big data have not yet taken place, one anticipates the emergence of such novel analytics in the near future. For instance, real-time analytics will likely become a prolific field of research because of the growth in location-aware social media and mobile apps. Since big data are noisy, highly interrelated, and unreliable, it will likely lead to the development of statistical techniques more readily apt for mining big data while remaining sensitive to the unique characteristics. Going beyond samples, additional valuable insights could be obtained from the massive volumes of less ‘trustworthy’ data.

ACKNOWLEDGMENT

The authors would like to acknowledge research and editing support provided by Ms. Ioana Moca.

REFERENCES

1. Aggarwal, C. C. (2011). An introduction to social network data analytics. In C. C. Aggarwal (Ed.), Social network data analytics (pp. 1–15). United States: Springer.

2. Barbier, G., & Liu, H. (2011). Data mining in social media. In C. C. Aggarwal (Ed.), Social network data analytics (pp. 327–352). United States: Springer.

3. Beaver, D., Kumar, S., Li, H. C., Sobel, J., & Vajgel, P. (2010). Finding a needle in haystack: Facebook's photo storage. In Proceedings of the nineth USENIX conference on operating systems design and implementation (pp. 1–8). Berkeley, CA, USA: USENIX Association.

4. Chen, H., Chiang, R. H. L., & Storey, V. C. (2012). Business intelligence and analytics: From big data to big impact. MIS Quarterly, 36(4), 1165–1188.

5. Chung, W. (2014). BizPro: Extracting and categorizing business intelligence factors from textual news articles. International Journal of Information Management, 34(2), 272–284.

6. Cukier K., The Economist, Data, data everywhere: A special report on managing information, 2010, February 25, Retrieved from http://www.economist. com/node/15557443

7. Diebold, F. X. (2012). A personal perspective on the origin(s) and development of "big data": The phenomenon, the term, and the discipline (Scholarly Paper No. ID 2202843). Social Science Research Network. Retrieved from http://papers.ssrn.com/sol3/papers.cfm?abstract id=2202843

8. Fan, J., Han, F., & Liu, H. (2014). Challenges of big data analysis. National Science Review, 1(2), 293–314.

9. Fan, J., & Lv, J.(2008). Sure independence screening for ultrahigh dimensionalfeature space. Journal of the Royal Statistical Society: Series B (Statistical Methodology), 70(5), 849–911.

10. Feldman, R. (2013). Techniques and applications for sentiment analysis. Communications of the ACM, 56(4), 82–89.

11. Gartner IT Glossary (n.d.). Retrieved from http://www.gartner.com/it-glossary/bigdata/

12. Gundecha, P., & Liu, H. (2012). Mining social media: A brief introduction. Tutorials in Operations Research, 1(4)

13. Hahn, U., & Mani, I. (2000). The challenges of automatic summarization. Computer, 33(11), 29–36.

14. Hakeem, A., Gupta, H., Kanaujia, A., Choe, T. E., Gunda, K., Scanlon, A., et al. (2012). Video analytics for business intelligence. In C. Shan, F. Porikli, T. Xiang, & S. Gong (Eds.), Video analytics for business intelligence (pp. 309–354). Berlin, Heidelberg: Springer.

15. Hasan, M. A., & Zaki, M. J. (2011). A survey of link prediction in social networks. In C. C. Aggarwal (Ed.), Social network data analytics (pp. 243–275). United States: Springer.

16. Heidemann, J., Klier, M., & Probst, F. (2012). Online social networks: A survey of a global phenomenon. Computer Networks, 56(18), 3866–3878.

17. He, W., Zha, S., & Li, L. (2013). Social media competitive analysis and text mining: A case study in the pizza industry. International Journal of Information Management, 33(3), 464–472.

18. Hirschberg, J., Hjalmarsson, A., & Elhadad, N. (2010). "You're as sick as you sound": Using computational approaches for modeling speaker state to gauge illness and recovery. In A. Neustein (Ed.), Advances in speech recognition (pp. 305–322). United States: Springer.

19. Hu, W., Xie, N., Li, L., Zeng, X., & Maybank, S.(2011). A survey on visual content-based video indexing and retrieval. IEEE Transactions on Systems, Man, and Cybernetics, Part C: Applications and Reviews, 41(6), 797–819.

20. Jiang, J. (2012). Information extraction from text. In C. C. Aggarwal, & C. Zhai (Eds.), Mining text data (pp. 11–41). United States: Springer.

21. Kwon, O., Lee, N., & Shin, B. (2014). Data quality management, data usage experience and acquisition intention of big data analytics. International Journal of Information Management, 34(3), 387–394.

22. Labrinidis, A., & Jagadish, H. V. (2012). Challenges and opportunities with big data. Proceedings of the VLDB Endowment, 5(12), 2032–2033.

23. Laney, D. (2001, February 6). 3-D data management: Controlling data volume, velocity and variety. Application Delivery Strategies by META Group Inc. Retrieved from http://blogs.gartner.com/doug-laney/files/2012/01/ad949-3D-DataManagement-Controlling-Data-Volume-Velocity-and-Variety.pdf

24. Liben-Nowell, D., & Kleinberg, J. (2003). The link prediction problem for social networks. In Proceedings of the twelfth international conference on information and knowledge management (pp. 556–559). New York, NY, USA: ACM.

25. Liu, B. (2012). Sentiment analysis and opinion mining. Synthesis Lectures on Human Language Technologies, 5(1), 1–167.

26. Manyika, J., Chui, M., Brown, B., Bughin, J., Dobbs, R., Roxburgh, C., et al. (2011). Big data: The next frontier for innovation, competition, and productivity. McKinsey Global Institute. Retrieved from http://www.citeulike.org/ group/18242/article/9341321

27. Pang, B., & Lee, L. (2008). Opinion mining and sentiment analysis. Foundations and Trends in Information Retrieval, 2(1–2), 1–135.

28. Panigrahi, B. K., Abraham, A., & Das, S. (2010). Computational intelligence in power engineering. Springer.

29. Parthasarathy, S., Ruan, Y., & Satuluri, V. (2011). Community discovery in social networks: Applications, methods and emerging trends. In C. C. Aggarwal (Ed.), Social network data analytics (pp. 79–113). United States: Springer.

30. Patil, H. A. (2010). "Cry baby": Using spectrographic analysis to assess neonatal health status from an infant's cry. In A. Neustein (Ed.), Advances in speech recognition (pp. 323–348). United States: Springer.

31. Schroeck, M., Shockley, R., Smart, J., Romero-Morales, D., & Tufano, P. (2012). Analytics: The real-world use of big data. How innovative enterprises extract value from uncertain data. IBM Institute for Business Value. Retrieved from http://www-03.ibm.com/systems/hu/resources/the real word use of big data.pdf

32. (2012, June 26). Small and midsize companies look to make big gains with "big data," according to recent poll conducted on behalf of SAP. Retrieved from http://global.sap.com/corporate-en/news.epx?PressID=19188

33. Sun, J., & Tang, J. (2011). A survey of models and algorithms for social influence analysis. In C. C. Aggarwal (Ed.), Social network data analytics (pp. 177–214). United States: Springer.

34. Tang, L., & Liu, H. (2010). Community detection and mining in social media. Synthesis Lectures on Data Mining and Knowledge Discovery, 2(1), 1–137.

35. TechAmerica Foundation's Federal Big Data Commission. (2012). Demystifying big data: A practical guide to transforming the business of Government. Retrieved from http://www.techamerica.org/Docs/fileManager.cfm?f=techamericabigdatareport-final.pdf

36. VanBoskirk, S., Overby, C. S., & Takvorian, S. (2011). US interactive marketing forecast 2011 to 2016. Forrester Research, Inc. Retrieved from https://www.forrester.com/US+Interactive+Marketing+Forecast+2011+To+ 2016/fulltext/-/E-RES59379

37. YouTube Statistics (n.d.). Retrieved from http://www.youtube.com/yt/press/ statistics.html

Chapter 14

Fusing Website Usability and Search Engine Optimization

Eugene B. Visser[1],Melius Weideman[2]

[1]*Purple Cow Communications, Cape Town, South Africa*

[2]*Website Attributes Research Centre, Cape Peninsula University of Technology, South Africa*

ABSTRACT

Background: Most websites, especially those with a commercial orientation, need a high ranking on a search engine for one or more keywords or phrases. The search engine optimisation process attempts to achieve this. Furthermore, website users expect easy navigation, interaction and transactional ability. The application of website usability principles attempts to achieve this. Ideally, designers should achieve both goals when they design websites.

Objectives: This research intended to establish a relationship between search engine optimisation and website usability in order to guide the industry. The authors found a discrepancy between the perceived roles of search engines and website usability.

Method: The authors designed three test websites. Each had different combinations of usability, visibility and other attributes. They recorded and analysed the conversions and financial spending on these experimental websites. Finally, they designed a model that fuses search engine optimisation and website usability.

Results: Initially, it seemed that website usability and search engine optimisation complemented each other. However, some contradictions between the two, based on content, keywords and their presentation, emerged. Industry experts do not acknowledge these contradictions, although they agree on the existence of the individual elements. The new model highlights the complementary and contradictory aspects.

Conclusion: The authors found no evidence of any previous empirical experimental results that could confirm or refute the role of the model. In the fast-paced world of competition between commercial websites, this adds value and originality to the websites of organisations whose websites play important roles.

INTRODUCTION

Background to the study

Small and medium-sized enterprises (SMEs) make up approximately 95% to 98% of all businesses in most countries. They provide many work opportunities and are essential for any country's economic growth (Samujh 2011). Information technology (IT) is also a main driver of economies. It facilitates the growth of SMEs through expansion into new markets, overcoming obstacles, allowing for quicker responses to changes in consumer patterns and allowing SMEs to compete internationally (Thurasamy *et al.* 2009). Internet-based technology is a significant part of IT and is an investment that drives innovation (Oliveira & Martins 2010).

In order for SMEs to sustain financial growth, they must adopt marketing systems that facilitate the buyers' and the sellers' decision-making processes (Layton 2011). In recent years, many SMEs have adapted to e-marketing. This allows dynamic business growth and changes the shape and nature of business by overcoming threats and creating new business opportunities (El-Gohary 2010). Because of e-business, e-marketing, e-commerce and internet user activity, SMEs often use websites to present and market products and/or services because the internet is rapidly becoming a communication, commerce and marketing medium that is changing business globally (Canavan, Henchion & O'Reilly 2007; Kűster & Vila 2011; Visser 2007).

After 1993, the release date of Archie Like Indexing for the WEB (ALIWEB), search engine development progressed rapidly. This occurred in the methods developers used to index information on the internet and the methods they used to provide searchers with the most relevant information available in the shortest time possible, given the search query.

The process of search engine optimisation (SEO) aims to improve the visibility of web pages to search engine crawlers. Today, SEO is an online marketing channel. SEO addresses design, architecture and content to allow for the better ranking of selected keywords. The results searchers receive are the search engine result pages (SERPs). They fall into two sections: organic (that occupies the primary real estate of the SERPs) and pay per click (PPC) that occupies the right-hand sides and sometimes the tops of the SERPs). Today, three major search engines dominate the market: Google, Bing and Yahoo! Of these, Google is by far the most popular (Carpineto *et al*. 2009). Currently, these three combined command over 95% of the world market in core searches per month (Adamo 2013).

In order to satisfy searchers' needs best, engines must extract and present the indexed information, which is most relevant to the searchers' queries, to the searchers for scrutiny. However, this is a complicated process because of the sheer volume of information available on the internet. Search engines must index this using search engine crawler programs and rank it appropriately using organic ranking algorithms.

The inner algorithmic workings of search engines are essential because searchers are often more interested in the quality of the search results than their quantity (Yang, Yang & Yuan 2007). Research has shown that, on average, searchers view no more than three SERPs for any particular search query. In fact, the closer any particular web page ranks to the first position on the first SERP, the higher the chances are for searchers to view that particular web page (Weideman 2009).

The research problem of this paper is that of the reduced productivity that results from the lack of guidance on the synergy between applying SEO and the usability of websites. The goal of this paper was to determine whether there is a relationship between search engine optimisation and website usability. If there is, this paper wants to determine this relationship. A model embodies this goal, which will be useful to industry and academe alike.

Literature review

Soon after search engines became popular, it emerged that people discovered between 42% and 86% of all websites through search engines (Thurow 2003). More recently, it emerged that almost 85% of all e-commerce began with queries submitted to search engines (Murphy & Kielgast 2008). These statistics make e-marketing a crucial component from a business perspective. Engines keep organic ranking algorithms confidential in order to avoid abuse (Jerkovic 2010).

An in-depth look at the literature on the search engine optimisation, website usability and spamdexing is necessary at this stage.

Search engine optimisation

SEO is the process of altering websites. It emphasises semantically themed keywords for search engines in order to improve website rankings. This, in turn improves websites' chances of being found in SERPs (Weideman 2009).

The head Google spam engineer has indicated that the Google organic ranking algorithm consists of approximately 200 elements. This suggests that Google considers 200 on- and off-page elements, which determine the relevancy of the search results presented in the SERPs (Cutts 2010). Many industry experts speculate what these elements are, along with the weight assigned to each element based on results when they conduct their own experiments. Several authors list a number of elements that would improve organic rankings (Google 2011d; Moz 2013; Sullivan 2011; Sullivan 2012).

The methodology uses understanding search queries from the searchers' perspectives as its basis. Search engines can attempt to determine the context of search queries, based on past searches, although this may seem almost impossible. Search engines do this by analysing the estimated time searchers spend on websites before returning to the search engines in order to submit refined queries and discussions on social networks that provide integrated and associated search engine results (Google 2011c; Rayson 2013). The foundation of the methodology uses the fact that search engines have interpretations of current indexed information as its basis. Therefore, search engines may actually be able to interpret the current indexed information better by associating search queries to the indexed content using searcher behaviour patterns. By monitoring ever-changing searcher behaviour patterns, search engines could constantly reorganise search results according to relevance (Google 2009; Sullivan 2009).

Google (along with other search engines) allows users (anonymous or identified) to share data with Google. Searchers conduct approximately 400 million searches every day on the Google search engine (Google 2011a; Enge *et al.* 2010). The vast amount of interpreted data search engines extract anonymously could give them enough information to make an appropriate interpretation of searched queries. Google's predictive search functionality is only one example of how Google uses searcher behaviour patterns based on popularity (see Figure 1).

Figure 1: Google's predictive search functionality.

Search engines also depend on editorial judgement, or link popularity, to interpret and determine the relevance of the indexed information better. Search engines would perceive a link from a credible website as a good quality link. Therefore, they would apply a positive 'vote' to the website at which the link is directed. Consequently, search engines can only perceive credible websites if the links they obtain come from other credible websites. This suggests that the quality of links they obtain is more important than the quantity of links. Furthermore, they need to associate the content on the website from which the link originated, and the content on the destination website of the same link, in some way from a thematic perspective. The anchor text (keywords the actual link uses) is also important and ideally should align with the destination web page's semantic themes and targeted keywords if search engines are to regard the link as worthy (Thurow 2008).

Search engines do yield relevant results. However, non-relevant results still appear in SERPs more often than not. In a perfect world, SEO marketers could assist search engines by ensuring that search engines apply only white hat SEO tactics to particular websites, thus 'cleaning' up the internet and reducing spamdexing. Unfortunately, as long as they can earn money through spamdexing, search engines will have to generate new ideas and organic ranking algorithm updates in order to discourage black hat SEO tactics. Eventually, this should reduce spamdexing.

Researchers have conducted new research recently. It involved university institutional repositories and using metadata to optimise them for search engine crawler visits. One study focused specifically on Google Scholar™. The researchers found that, if they changed the metadata in these repositories from Dublin Core to Google Scholar's own prescriptions, indexing rose significantly (Arlitsch & O'Brien 2012).

SEO is an online marketing strategy. The basis of the SEO strategy is to address website design, architecture and content so that search engine spiders can make appropriate interpretations in terms of the themed keywords and rank websites accordingly. Strictly speaking, the goal of SEO is to satisfy search engine spiders and organic ranking algorithms by aligning the targeted themed web page content with the semantically interpreted keywords when considering search queries.

Spamdexing

The basis of the search engine revenue model is PPC advertising. However, the search functionality is available at all times and is free for public use. Search engine success still depends on providing relevant results to searchers in the shortest time possible as non-relevant and/or slow results may deter searchers from using particular search engines. Therefore, non-relevant results could have a direct effect on search engines' revenue.

Because of the direct association between search engine rankings and searchers' behaviour when viewing SERPs, it becomes clear that website owners may attempt to manipulate search engine rankings to increase their businesses' exposure. Spamdexing (also known as search engine spam) refers to websites that attempt to deceive search engines, whereby the results they provide to searchers are not relevant when one considers their search queries (Weideman 2009). Search engines regard this behaviour as unscrupulous and unsolicited.

As a result, major search engines use engineers who focus only on eliminating spamdexing. However, search engines do not publish their preventative spamdexing rules. Instead, they publish their (mostly vague) best practice guidelines. They do this to protect their organic ranking algorithms because are part of the foundation of their revenue models (Enge *et al.*2010). At the same time, some search engines regularly adapt their algorithms to detect and flag websites that contain attempts at spamdexing. In Google's case, some of these updates (termed Panda and Penguin) often have dramatic effects on a small percentage of search results (Cooper 2013; Quinton 2012). The industry watches these changes closely. They normally cause a spurt of activity because commercial concerns could experience decreases in revenue if their websites drop in position on the ranking lists.

Spamdexing can occur on two levels: content manipulation and link structure manipulation. Content manipulation is restricted to on-page factors – elements over which only website authors have control. In earlier days, search engines depended primarily on what search engine spiders could see during indexing. This allowed website authors to manipulate web page structure so that search engine spiders could not see some content and only visitors could – and vice versa. Furthermore, it often included targeted keywords and/or repeated them a number of times on a particular web page (keyword stuffing), whereby the web page content would not make contextual sense to visitors. However, search engine spiders would prioritise the keywords irrespective of the website's semantic theme for ranking purposes (Wu & Davison 2006). This is becoming less of a problem today because search engines' algorithms now depend more on link popularity. Conversely, this discloses the opportunity to link structure manipulation. Because of the importance of link popularity, in terms of organic ranking algorithms, one should note that link popularity manipulation is more difficult to detect than is content manipulation.

In the past, black hat SEO marketers sometimes sold apparently undetectable spam methods. However, Google engineers have indicated that one can detect spamdexing (Cutts 2007). This is an interesting statement given the recent findings of spamdexing in the Google SERPs because of link manipulation. The JC Penney fiasco provided more than enough evidence to indicate that one can regard Google's best practice guidelines for link schemes as nothing more than interesting reading (Google 2011b). An article, which the *New York Times* published, discovered that JC Penney (a department store chain) obtained over 2000 links, mostly from non-related websites. They varied from nuclear engineering and property portholes to casino-focused websites, all with the appropriate JC Penney themed anchor text (see Segal 2011). Although the JC Penney website seemed to have participated in search engine prohibited link schemes, the website still maintained high rankings for a number of targeted terms in Google SERPs. This discovery convinced SEO marketers that link farms and off-topic website linking schemes still have positive ranking effects, despite Google best practice guidelines. The recent Google organic ranking algorithm update 'Panda' (as a direct result of the JC Penney fiasco) is another attempt to fight spamdexing and enforce prohibited link schemes.

Spamdexing is increasing significantly, simply because black hat SEO marketers can make money from it. Unfortunately, it is difficult to measure search quality because search result quality is relative to searchers' perspectives of their search queries. Ultimately, automated algorithmic solutions may not be the best way to solve the spamdexing problem. The different search engines may actually have to collaborate in order to find an effective solution. However, it is interesting to note that the Google search engine may be responsible, to some extent, for a number of spamdexing websites. Personally owned websites, as opposed to SERPs, display Google AdSense™ ads (which is part of PPC advertising). Websites that contain

AdSense™ ads are, in some circumstances, created with the sole purpose of obtaining visitors who could click on any of the displayed AdSense™ ads. With every click on any of the AdSense™ ads, Google earns money, as does the author of the website that displays the AdSense™ ads. Therefore, certain search engines might first have to look at their revenue models and their functionality before addressing the spamdexing problem successfully (Cutts 2011).

Website usability

Visitors to a website usually have specific questions about a particular problem or need (Eisenberg *et al*. 2008). Although one may perceive search engines as facilitators, they must still address searchers' needs in terms of finding the appropriate website associated with the search queries. Searchers have come to understand that search engines attempt to provide the most relevant results first. This implies that, if they do not obtain the 'correct' result within the first three SERPs, the remaining results will probably also be irrelevant. Because search engines do not always comprehend search queries from the searchers' perspectives fully – often resulting in non-relevant and/or spamdexing results – searchers must frequently alter their search queries to clarify the information they need.

Figure 2 illustrates a typical searcher's process for finding information.

Searchers seldom leave search engines if they do not meet the searchers' information needs. The reason for this is that search engines are the facilitators and because there are not many alternatives (in layout and/or functionality) that can guarantee that they will meet the information needs. On the other hand, visitors treat websites differently.

One can categorise the intentions and motivations for visiting websites as:

- exploration
- information
- entertainment
- shopping.

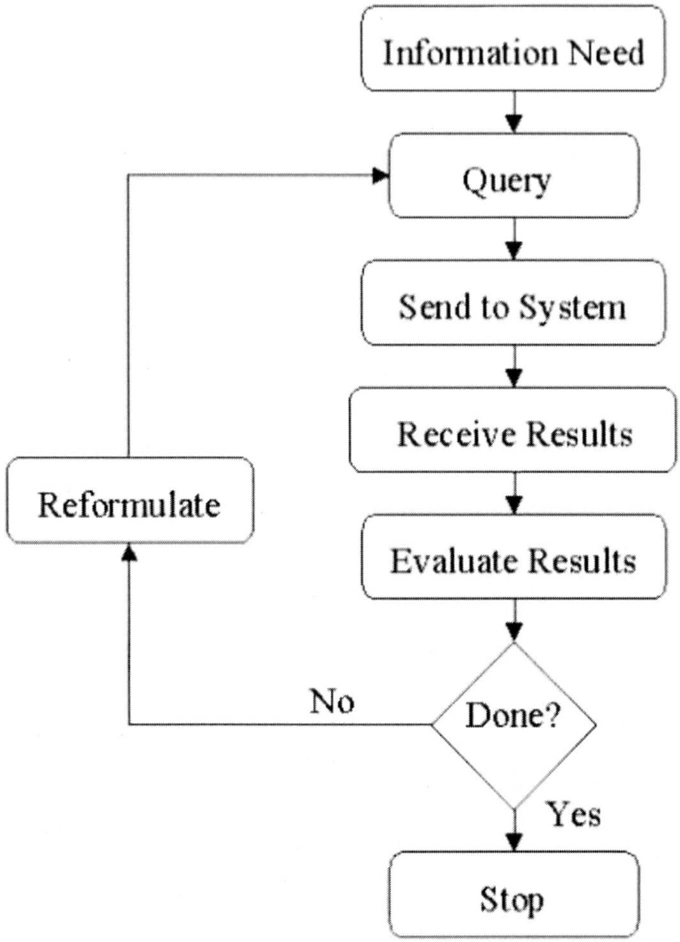

Figure 2: Standard information access process.

Whatever reasons visitors have for visiting websites, developers create websites for information, opinion, marketing and/or for financial gain in one way or another. Website usability (WU) addresses the functional application of information about visitors' ability to interact successfully with that information. Therefore, the goal is to remove any obstacles that could impede visitors' experiences when interacting with websites (Eisenberg *et al.* 2008).

Website usability defines the quality of the visitors' experiences because it addresses mechanical and persuasive usability problems (Visser & Weideman 2011a; Visser & Weideman 2011b). Website usability statistics have shown that there is only a 12% probability that visitors will revisit a particular website. This shows that, once a website has lost a visitor because of the lack of WU, the visitor is usually lost for good (Nielsen & Loranger 2006). Therefore, one should see WU as important (if not more important) than SEO, which, in turn, emphasises the importance of fusing SEO and WU.

Although developers ultimately create websites for visitors, the visitors may initially not be aware of the websites' existence. Searchers are never fully aware of all the websites on the internet relevant to particular search queries. Therefore, search engines that will provide searchers with the most relevant results using organic ranking algorithms' interpretations are necessary. Although designers create these search engine results dynamically, the results are still subjective. They depend on the organic ranking algorithms that often provide irrelevant and/or spam results.

In addition, the SERPs may not have listed the ideal website results (or listed them very low down), because those websites were not visible to the search engine spiders (or did not satisfy the organic ranking algorithms appropriately). Search engine spiders and organic ranking algorithms depend on a number of pre-programmed rules that correspond with the conceptual models of the internet. This shows that, if search engine spiders have not crawled and indexed a website, that website cannot possibly rank in the SERPs. Furthermore, if the organic ranking algorithms do not interpret the web page appropriately, then that web page may not rank at all for the targeted keywords.

In summary, website authors must regard the search engine optimisation guidelines (SEO elements) as priorities whilst designing and developing websites if search engines are to crawl, index and rank websites for targeted keywords.

One should not interpret WU as the usefulness of websites from personal angles. Instead, one should regard WU as a task-orientated function from anonymous, yet personal perspectives. It is important for visitors to know, exactly and intuitively, how to accomplish specific tasks on any particular website. Therefore, WU addresses the effectiveness, efficiency, learnability, memorability, error recovery and satisfaction of websites (Thurow & Musica 2009). Although artificial intelligence or algorithms do not govern it, WU has a number of attributes that one could apply to particular websites in order to improve interaction. All websites are different. However, the fundamental attributes remain the same in terms of how visitors interpret web pages. The actual design of websites is a crucial component of WU, and, by using graphic and textual signals, visitors should, at any point, be able to identify the current location, as well as the process involved in order to reach the desired destination whilst considering their objectives.

SEO is an essential component, as is WU, of successful websites. Therefore, SEO and WU will form the foundation for successful websites. However, spamdexing is also an essential component because developers must apply SEO so that search engines ideally never perceive business websites as spam. In addition, information-seeking platforms are changing. This will have an effect on WU if websites do not change to accommodate platform-specific devices appropriately (Nicholas*et al*. 2013). One can view successful SEO as the card to draw visitors to websites (via SERPs), whilst usability is the glue that keeps visitors interacting with websites. When combined correctly, the two will convert searchers to users to buyers – an action often called conversion.

Research method and design

The effect of SEO on websites depends largely on a simultaneous combination and deployment of all SEO elements. On the other hand, WU addresses the functional application of information to visitors.

The authors used a pre-test and post-test quantitative methodological design. They analysed the on-page SEO elements, which conflict with WU attributes, by using three websites that offer identical services. The authors analysed:

- **The control website (CW):** The CW (which a business with minimal knowledge of SEO and WU created) has existed since 2006 (http://www.copywriters.co.za). It consists of 34 pages and 17 114 words of content. Because of the domain age, search engines had already successfully crawled and indexed the website. In addition, the website had generated a number of existing inlinks from several sources on the world wide web. However, for the purpose of this experiment, the authors excluded the referrer and direct traffic sources from the traffic-source data collection. One should note that the authors made no changes to the CW during the experiment.

- **The experimental website (EW):** The authors created the EW with WU in mind. However, they deliberately ignored all SEO elements with minimal content (http://www.copywriters.co.za/ppc/). They isolated it from all forms of website traffic other than that PPC generated. They obtained user feedback by measuring the number of conversions the website obtained. One should also note that they made no changes to the EW during the experiment.

- **Experimental website 2 (EW2):** The authors created EW2 with SEO in mind. However, they deliberately ignored all WU attributes. They launched the newly formed http://www.translation-copywriters.co.za website on 01 July 2010. It had no existing in-links, 29 web pages and 48 923 words of content. They also excluded referrer and direct traffic from the traffic-source data collection on this website. Organic and PPC traffic were the primary traffic sources. The first search engine crawled and indexed this website on 08 July 2010. During the next four months, the authors made systematic SEO changes to the website and recorded the primary experiment ranking measurements on 08 November 2010. These changes appear below.

    - On 08 August 2010, the authors added 8077 words. They also optimised the metadata. This included headings and the anchor text in terms of the keywords.

    - On 08 September 2010, the authors added 5362 words. They also optimised the metadata. This included headings and the anchor text in terms of the keywords (similar to the changes the authors made in the previous month).

    - On 08 October 2010, the authors added 5726 words. They applied theming and internal linking to emphasise contextual phrasing and semantically related keywords.

    - On 08 November 2010, the authors added 3080 words. They increased keyword density, frequency and proximity using the benchmarks they obtained for high-ranking competitors on the same-targeted keywords.

The primary experiment focused on monitoring 130 specific keywords and their rankings on the three major search engines (Google, Yahoo! and Bing). The intention was to compare the control website's rankings with those that Experimental Website Two generated in order to determine whether applying SEO elements to a particular website had a direct effect on improved website ranking for targeted keywords. An extract of this keyword or phrase list follows.

Ad jingles; advertising and editorial writing; advertising copy; industry specific articles copywriting; informative media writing; internet copywriting; jingle writing; strategic business copywriting; strategic copywriting; technical editing; translate English into Afrikaans.

The authors conducted three additional experiments:

- **Organic traffic:** The authors measured organic traffic for the CW and the EW2 daily on each domain. They grouped the results every month and interpreted them using a linear regression analysis. The objective was to determine whether the organic ranking improvements had a direct effect on increases in organic traffic.

- **Conversion (submitting a contact form):** To establish the effectiveness of WU, the authors conducted conversion testing using a PPC campaign that they applied to all three websites for 49 days, with a budget of R3000.00 each. They decided to use the Kruskal-Wallis test to inspect the difference between variables. The specific variables the authors examined were the *average time on site per visitor*, *the average page views per visit* and *the average number of conversions obtained per visitor*. The objective was to determine whether the WU attributes, which contradict SEO elements, are essential to implement in order to improve website conversions.

- **Interviews:** The authors conducted interviews with five randomly selected website users with a minimum of eight years of internet exposure in order to:

    1. Identify WU attributes that they might have overlooked.

    2. Consider human interaction compared to a focus on theory only.

RESULTS

Through statistical analysis (using a univariate analysis of variance testing), the authors determined whether or not the web page to which a search was directed and the direction from which the search originated (global or local) has a significant effect on ranking positions. Although the results from the primary experiment, recorded on 08 November 2010, indicated that the search engine from which the search originated does not have a significant effect on rankings, the authors determined that, in all instances, the rankings for the EW2 were significantly better than those of the CW were. This showed that applying SEO elements to a particular website had a direct effect on improved website rankings for targeted keywords.

The additional experiments the authors conducted revealed:

- **Organic traffic:** Using linear regression analysis, the authors determined that the coefficient for X showed that the traffic trend for the EW2 increased significantly over time. However, the coefficient for X on the Control Website indicated there was no significant trend for traffic over time. The authors determined that 51.4% of the variation in traffic to both websites was because of changes in time and alterations to the website. This showed that the organic ranking improvements had a direct effect on organic traffic increases.

- **Conversion:** The results the authors gleaned from the conversion analysis of all three websites on the three main variables (*average time on site per visitor*, *the average page views per visit* and *the average number of conversions obtained per visitor*) showed that there was a significant statistical difference between the CW, EW and EW2. The EW yielded the highest number of conversions. This shows that applying WU attributes is essential to improving website conversions.

- **Interviews:** The user feedback showed that WU attributes are important factors when visitors are deciding whether or not to interact with a website. The authors determined that participants perceived that some SEO elements were obstacles to WU. This confirmed the existence of SEO and WU contradictions. The participants agreed unanimously that the EW website was the one that provided the best user experience.

On the surface, it seems that SEO and WU complement each other (Visser & Weideman 2011b). However, one should note that SEO and WU contradict each other in some cases. Figure 3 presents the 'fused SEO and WU model' on the left (marked A). On the right, marked B, Figure 3 presents the key to interpreting the model. The model consists of three main sections: SEO, WU and additional considerations. The connecting red lines illustrate the contradictions between SEO and WU. The bottom of the Figure illustrates the fused SEO and WU solution.

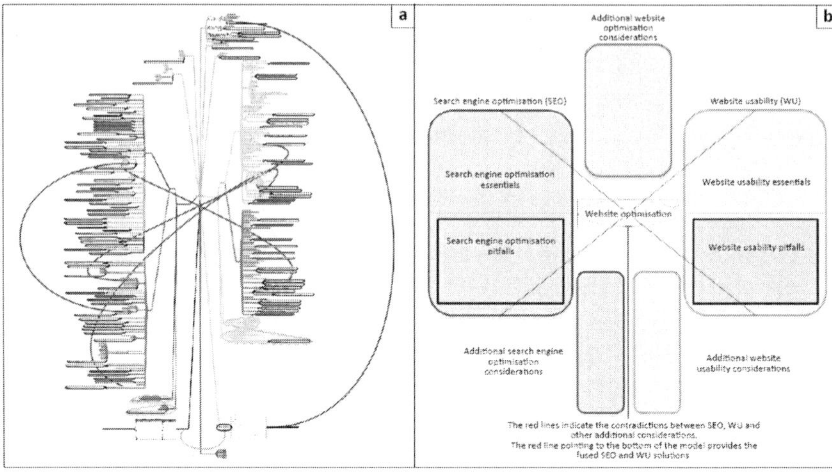

Figure 3: The fused search engine optimisation and website usability model (a) and the fused search engine optimisation and website usability model key and colour codes (b). (A scalable image of the model is available at http://www.eugene-visser.co.za)

Essentially, the contradictions between SEO and WU revolve around content, keywords and their presentation. Search engines are not human and, regardless of how advanced artificial intelligence may become, the probability of their completely simulating human behaviour is low. Essentially, search engine crawlers consider two components: the information actual web pages provide and other web pages that give opinions about the information on the web pages. The opinions of others could have an effect on the human decision-making process. However, reflecting on previous actions and human intuition defines the human decision-making process (Eisenberg *et al.* 2008; Pather & Remenyi 2005).

Furthermore, search engines also need to evaluate the competitive component for ranking priorities. This shows that it is important to understand what websites are about and whether website X will better satisfy visitors' needs compared to website Y. In order to achieve this, websites must emphasise the information they provide to search engines. This would lead to an enormous amount of content and keyword or phrase semantic emphasis. Conversely, visitors are not interested in an overwhelming amount of content. This could entice visitors to leave websites out of frustration.

Other contradictions revolve around how search engines have attempted to simulate visitors' interpretations of websites by evaluating their behaviour patterns. Search engines assume that, if a visitor visits a single web page on any particular website and leaves shortly afterwards, the web page did not satisfy the visitor's needs. This are 'bounces' and have negative connotations. The contradiction exists with landing pages and conversion optimisation, where search engines optimise web pages to provide all the necessary information and functionality to satisfy visitors' needs best given their objectives.

Ironically, industry experts do not acknowledge any contradictions. However, in both SEO (Thurow & Musica 2009) and WU (Nielsen & Loranger 2006:16), experts indicate that there are elements and attributes, which occur during website optimisation, that do not consider competitors. When scrutinising the SEO elements and WU attributes, which the industry experts define, the model reveals the contradictions (see Figure 3).

Developers could resolve the identified contradictions in the website architecture whilst developing websites. The solution uses SEO methodology, whereby developers categorise websites into themes, thus isolating each category within the website for emphasis (to address the competitive component).

Ideally, each category should consist of a number of content heavy web pages based on semantically targeted keywords or phrases (in order to address search engine phrase indexing). Therefore, the internal linking structure within each category needs to emphasise the primary category web page (landing page) by linking the actual semantically-related keywords or phrases in content web pages to the landing page (in order to address semantic-related keyword emphasis). Editorial judgement (inbound links) should target the appropriate semantic phrase content pages. This, in turn, will emphasise the landing pages. The high authority (good quality) websites relevant to the category should link directly to the landing page. Therefore, primary navigation should consist of the landing pages (which should be the actual product or service pages). Ideally, developers should optimise landing pages in order to optimise conversion. This will provide all the necessary information and functionality to allow visitors to convert on the landing pages.

Applying this architectural methodology (along with all other SEO elements) should address the SEO requirements of websites.

DISCUSSION AND CONCLUSIONS

The SEO methodology prioritised the landing pages during search engine ranking. This indicates that visitors will arrive on the appropriate category web page, which should align with the visitors' search queries (in order to address the visitors' particular needs). Here, the primary navigation and using the breadcrumbs would already have addressed the WU in terms of current location and desired destination.

Optimising the landing pages for conversion will reduce unnecessary clicks and visitor frustration. Reducing web page content is a major challenge. However, developers can reduce content through the functionality of expendables. Developers should only apply this solution where appropriate and with caution. The functionality entices website interaction, allowing visitors to request additional information on the same web page without impeding interaction and/or visitor experience.

Although a few search engines use this technology, developers must implement it correctly to ensure that crawlers will crawl and index the content of entire web pages.

In conclusion, a listing in the top position on SERPs is not enough. In addition to the web page result description being enticing enough to convince searchers to click on the results (assuming that the search results align with the search queries), it is essential that searchers engage with websites in terms of their objectives (Thurow 2008; Visser & Weideman 2011b).

Finally, unforeseen contradictions may still surface during website development, even when one considers the identified SEO elements, WU attributes and the suggested website architecture.

The authors created the model in Figure 3 to provide a website optimisation guide for any business. It will address SEO and WU simultaneously. This model is the major contribution this research has produced to knowledge.

The reality is that each website is unique and its developer must optimise it with critical considerations of its business objectives.

COMPETING INTERESTS

The authors declare that they have no financial or personal relationship(s) that may have inappropriately influenced them when they wrote this article.

AUTHORS' CONTRIBUTIONS

E.B.V. (Purple Cow Communications) was the doctoral student, was responsible for the experimental and project design and producing the first draft of the manuscript. E.B.V. also effected ongoing changes and improvements and also created and managed the test and result websites. M.W. (Cape Peninsula University of Technology) initiated this research, was the only supervisor, contributed ideas on an ongoing basis to the design and experimentation, selected the journal, did the proofreading and layout and handled all editorial matters. Both authors made conceptual contributions to this research, participated in the writing and approval of this research project.

REFERENCES

1. Adamo, S., 2013, 'comScore releases August 2013 U.S. Search Engine Rankings', in *comScore*, viewed 16 October 2013, from http://www.comscore.com/Insights/Press_Releases/2013/9/comScore_Releases_August_2013_U.S._Search_Engine_Rankings

2. Arlitsch, K. & O'Brien, P.S., 2012, 'Invisible institutional repositories: Addressing the low indexing ratios of IRs in Google Scholar', *Library Hi Tech* 30(1), 60–81.

3. Baeza-Yates, R. & Ribeiro-Neto, B., 1999, *Modern information retrieval*, Addison-Wesley, Harley.

4. Canavan, O., Henchion, M. & O'Reilly, S., 2007, 'The use of the Internet as a marketing channel for Irish speciality food',*International Journal of Retail & Distribution Management* 35(2), 178–195.

5. Carpineto, C., Osiński, S., Romano, G. & Weiss, D., 2009, 'A survey of web clustering engines', *ACM Computing Surveys*41(3), 17.1–17.38.

6. Cooper, L., 2013, 'Aiming high', *Marketing Week* 41(3), 39–41.

7. Cutts, M., 2007, 'Undetectable spam', in *Matt Cutts: Gadgets, Google, and SEO*, viewed 01 May 2013, fromhttp://www.mattcutts.com/blog/undetectable-spam/

8. Cutts, M., 2010, 'Google incorporating site speed in search rankings', in *Matt Cutts: Gadgets, Google, and SEO*, viewed 01 February 2013, from http://www.mattcutts.com/blog/site-speed/

9. Cutts, M., 2011, 'My thoughts on this week's debate', n *Matt Cutts: Gadgets, Google, and* SEO, viewed 26 January 2013, from http://www.mattcutts.com/blog/google-bing/

10. Eisenberg, B., Quarto-von Tivadar, J., Davis, L.T. & Crosby, B., 2008, *Always be testing: the complete guide to Google website optimizer*, Sybex, Indianapolis.

11. El-Gohary, H., 2010, 'E-Marketing – A literature review from a small business perspective', *International Journal of Business and Social Science* 1(1), 214–244.

12. Enge, E., Spencer, S., Fishkin, R. & Stricchiola, J.C., 2010, *The art of SEO – Mastering search engine optimization*, O'Reilly Media Inc., Sebastopol.

13. Google, 2009, *Personalized search for everyone*, viewed 25 August 2013, fromhttp://googleblog.blogspot.com/2009/12/personalized-search-for-everyone.html

14. Google, 2011a, *Frequently asked questions for the Google analytics data sharing options*, viewed 29 November 2012, fromhttp://www.google.com/support/analytics/bin/answer.py?answer=87515

15. Google, 2011b, *Link schemes*, viewed 27 January 2013, from http://www.google.com/support/webmasters/bin/answer.py?answer=66356

16. Google, 2011c, *High-quality sites algorithm goes global, incorporates user feedback*, viewed 25 August 2013, fromhttp://googlewebmastercentral.blogspot.com/2011/04/high-quality-sites-algorithm-goes.html

17. Google, 2011d, *More guidance on building high-quality sites*, viewed 25 August 2013, fromhttp://googlewebmastercentral.blogspot.com/2011/05/more-guidance-on-building-high-quality.html

18. Jerkovic, J.I., 2010, *SEO warrior*, O'Reilly Media Inc., Sebastopol.

19. Kűster, I. & Vila, N., 2011, 'Successful SME web design through consumer focus groups', *International Journal of Quality & Reliability Management* 28(2), 132–154.

20. Layton, R.A., 2011, 'Towards a theory of marketing systems', *European Journal of Marketing* 45(1/2), 259–276.

21. Moz, 2013, *Google algorithm change history*, viewed available 25 August 2013, from http://moz.com/google-algorithm-change

22. Murphy, H.C. & Kielgast, C.D., 2008, 'Do small and medium-sized hotels exploit search engine marketing?', *International Journal of Contemporary Hospitality Management* 20(1), 90–97.

23. Nielsen, J. & Loranger, H., 2006, *Prioritizing web usability*, New Riders Press, Berkeley.

24. Nicholas, D., Clark, D., Rowlands, I. & Jamali, H.R., 2013, 'Information on the Go: A case study of Europeana mobile users', *Journal of the American Society for Information Science and Technology* 64(7), 1311–

25. Oliveira, T. & Martins, M.F., 2010, 'Understanding e-business adoption across industries in European countries', *Industrial Management & Data Systems* 110(9), 1337–1354.

26. Pather, S. & Remenyi, D., 2005, 'Some of the philosophical issues underpinning research in information systems – From positivism to critical realism', *South African Computer Journal* 35, 76–83.

27. Quinton, B.D., 2012, 'Google's Penguin update: Unhappy feat for marketers?', *Chief Marketer* 3(10), 8.

28. Rayson, S., 2013, *10 Ways Google+ will improve your SEO*, in *socialmediatoday*, viewed 25 August 2013, from

29. Samujh, H., 2011, 'Micro-businesses need support: survival precedes sustainability', *Corporate Governance* 11(1), 15–28.

30. Segal, D., 2011, 'The dirty little secrets of search', in *The New York Times*, viewed 02 February 2013, fromhttp://www.nytimes.com/2011/02/13/business/13search.htm l

31. Sullivan, D., 2009, *Google now personalizes everyone's search result*, in *Search Engine Land*, viewed 25 August 2013, fromhttp://searchengineland.com/google-now-personalizes-everyones-search-results-31195

32. Sullivan, D., 2011, *Why Google Panda is more a ranking factor than algorithm update*, in *Search Engine Land*, viewed 25 August 2013, from http://searchengineland.com/why-google-panda-is-more-a-ranking-factor-than-algorithm-update-82564

33. Sullivan, D., 2012, *Google Penguim update recovery tips & advice*, in *Search Engine Land*, viewed 25 August 2013, fromhttp://searchengineland.com/penguin-update-recovery-tips-advice-119650

34. Thurasamy, R., Mohamad, O., Omar, A. & Marimuthu, M., 2009, 'Technology adoption among small and medium enterprises (SME's): A research agenda', *World Academy of Science, Engineering and Technology* 53, 943–946.

35. Thurow, S., 2003, *Search engine visibility*, New Riders Press, Indianapolis.

36. Thurow, S., 2008, *Search engine visibility*, 2nd edn., New Riders Press, Indianapolis. PMCid:PMC2613660

37. Thurow, S. & Musica, N., 2009, *When search meets web usability*, New Riders Press, Berkeley. PMCid:PMC2676734

38. Visser, E.B., 2007, 'Search engine optimisation elements' effect on website visibility: The Western Cape real estate SMME sector', unpublished MTech thesis, Cape Peninsula University of Technology.

39. Visser, E.B. & Weideman, M., 2011a, An empirical study on website usability elements and how they affect search engine optimisation, *SA Journal of Information Management* 13(1), Art. #428, 9 pages.

40. Visser, E.B. & Weideman, M., 2011b,' Search engine optimisation versus Website usability – Conflicting requirements?',*Information Research* 16(3), paper 493.

41. Weideman, M., 2009, *Website visibility: The theory and practice of improving ranking,* Chandos Publishers, Oxford.

42. Wu, B. & Davison, B.D., 2006, Detecting semantic cloaking on the Web, in *Proceedings of the International World Wide Web Conference Committee (IW3C2)*, Edinburgh, Scotland, 23–26.

43. Yang, C., Yang, K. & Yuan, H., 2007, 'Improving the search process through ontology-based adaptive semantic search',*The Electronic Library* 25(2), 234–248.

Chapter 15

Use of Social Media Platforms for Improving Academic Performance at Further Education and Training Colleges

Godwin P. Dzvapatsva[1], Zoran Mitrovic[1], Anthony D. Dietrich[1]

[1]*Department of Information Systems, University of the Western Cape, South Africa*

ABSTRACT

Background: The National Certificate Vocational (NC[V]) curriculum offered by Further Education and Training (FET) colleges was introduced in 2007 to address the skills shortage in South Africa. Information Technology (IT) lecturers encountered a number of challenges in delivering lessons throughout the course, which affected the academic performance of learners. The biggest challenges identified were the lack of adequate contact hours for the course and inconsistency in the way in which final examination papers were set.

Objectives: The aim of the project was to investigate the use of: (1) a knowledge portal for verifying the quality of assessments by lecturers and (2) social media to increase contact time with FET college students in an attempt to improve their academic performance.

Method: The NC(V) level 3 student test scores for 2011 were compared to those of 2012. In addition to the test scores, students also received a questionnaire so as to determine their perceptions on social media usage.

Lecturers also received a questionnaire on their perception of the knowledge portal.

Results: The data collected from seven lecturers and 38 students indicated a 35% (from 30% – 65%) improvement in academic performance after the introduction of the interventions, that is social media and a knowledge portal; an indication of the importance of electronic media in enhancing learning.

Conclusion: The research offered FET lecturers an additional method for learning and teaching in that they could use the knowledge portal to set up quality assessments for the students and social media to increase contact learning time.

INTRODUCTION

The skills shortage in South Africa has seen the government making frantic efforts to address the problem through Further Education and Training (FET) colleges. One such initiative was the National Certificate Vocational (NC[V]) curriculum, which was introduced in 2007. The target population for the NC(V) curriculum is learners who passed Grade 9. As reflected by the low pass rate of 2011 and preceding years, this curriculum is however proving to be challenging to most of the students. It remains the task of the lecturer to come up with innovative ways to ensure that students perform well. This study suggests the use of social media (SM) by students and a knowledge portal (KP) for assessments by lecturers to improve academic efficiency. The two suggested solutions, SM and a KP, have been found in separate studies to be vital in the sharing of knowledge between students and lecturers (Dietrich, Whyte & Mitrovic 2011; Dzvapatsva 2013; Koles & Nagy 2012). All the knowledge generated from previous lecturers and other sources, for example departmental sample papers, is stored and transferred to other lecturers through the KP for assessments. The students are encouraged to use the Facebook (FB) group created by lecturers to increase contact time outside normal learning time.

It is presumed that most of the senior staff at the FET colleges grasp the need to share knowledge with the junior staff so as to increase or enhance learning; thus, they 'are eager to introduce knowledge management paradigms' (Bock & Kim 2002). From the literature (Koles & Nagy 2012) it appears that SM platforms have a remarkable effect on learning.

The use of electronic platforms, especially SM, for the purpose of knowledge sharing amongst FET students has been criticised by various researchers (Hamid, Chang & Kurnia 2009; Sarachan & Reinson 2011). A random sample of lecturers at a South African FET college indicated that most of the students from non-IT departments were against the idea of using electronic platforms, especially SM, for the purpose of knowledge dissemination amongst students and lecturers. These lecturers argued that electronic platforms like FB are disruptive and destructive. However, other researchers (Kabilan, Ahmad & Abidin 2010) found the opposite to be true. They argued that SM platforms like FB could bring positive results in academic performance, if well administered. This is why the authors of this article decided to use FB to increase the contact time required for the subject by the students and the KP for knowledge sharing amongst lecturers. It was the view of these researchers (Dzvapatsva 2013; Dietrich *et al.* 2011) that NC(V) was heavily congested in terms of the time needed for the subject and students were left with not enough time to study or to do research whilst at the college (Leung 2002). Owing to the geographical distance, students could be contacted via SM platforms, like FB, which could accessed by most students from any location.

Apart from the students, the lecturers do not have enough time to do research or examine the subject content from previous lecturers or fellow subject experts. The KP seems to be the best electronic source which lecturers can use as reference for work done by previous and current subject experts. At the same time, FB seems to offer an alternative in capturing students' attention outside the classroom to review what they have learnt or will learn in the following lecture.

BACKGROUND

The NC(V) programme was introduced to FET colleges in 2007 in an effort to redress the skills imbalance in South Africa. All learners who wanted to proceed with their studies along a chosen career path, for example Information Technology (IT), Business Administration, Engineering, Marketing or Tourism, could opt for the NC(V) programme. The NC(V) programme draws its students from Grade 9–12. The entry qualification is a pass in Grade 9, or the student must be older than 16 years. It runs for a period of three years, thus level 2 to level 4. Each year is an exit point. At each level a student is expected to successfully complete seven subjects, which comprise four core subjects, two fundamental ones, and one elective. Most students, however, struggle to pass all seven subjects so as to be promoted to the next level. The promotional policy allows students with a pass in five subjects to proceed to the next level. Lecturers have to work extremely hard and implement measures to improve students' performance so that FET colleges remain economically viable; these measures include the use of electronic platforms. In the event that a student is promoted to the next level whilst requiring one or two subjects for the previous level, he or she will have to do seven subjects of the present level and two extra subjects, which becomes stressful for the student. In view of the aforementioned, lecturers had to design means to assist students and to improve academic efficiency at each level, hence the introduction of electronic platforms. At the South African FET colleges used for this study, the SM platform, Facebook, has been used to contact learners outside the normal teaching time to increase contact hours for the Programming component, which has been identified as one of the challenging subjects. The lecturers also used a portal for the setting up of assessments, subject notes, tests, and worked-out solutions. This has presumably been suggested to assist in drastically reducing the bottleneck and directly improve the throughput rate of students at FET colleges, which ultimately will contribute towards overall academic fulfilment (Dzvapatsva 2013). Studies by researchers, including Kabilan *et al*. (2010), has shown the positive use of SM for academic improvement, ultimately improving the throughput rate. In parallel, researchers like Rajalakshmi and WahidiBanu (2009) identified the KP application as a possible solution to academic inefficiency.

LITERATURE REVIEW

It should be noted that whilst the traditional medium of face-to-face interaction offers numerous advantages for teaching and learning, it is controlled by time and space. In that respect it becomes crucial to embrace new technologies to supplement the current traditional ones in the teaching and learning of college and university students. With electronic media, the world becomes the classroom, available 24/7, and not confined to Mondays to Fridays.

According to Kaplan and Haenlein (2010), SM is defined as 'a group of Internet-based applications that build on the ideological and technological foundations of Web 2.0, and that allow the creation and exchange of User Generated Content'. The two main aspects discussed here are media research and social processes. Each medium used by the lecturer differs in its degree of effectiveness, depending on whether it is visual, acoustic or physical (Short, Williams & Christie 1976). Taking this into consideration, the current generation of students are more visual, hence the use of FB. It should also be considered that these students are a product of Outcome Based Education (OBE), which requires them to demonstrate the skills and subject content that they are expected to learn, regardless of whether they are comfortable with the medium used. Students prefer a medium which increases self-disclosure and self-presentation, thereby increasing social influence. The more a medium reduces ambiguity and uncertainty, the more it is accepted by communicating students (Daft & Lengel 1986).

Generally, SM encompasses the following:

- social networking sites (SNS), such as Facebook, Twitter, MySpace and LinkedIn
- media sharing sites, such as YouTube, Flickr and Tumbler
- wikis and blogs
- syndication of content through Rich Site Summary (RSS) feeds
- republishing tools (Gikas & Grant 2013).

According to Greenhow (2011, cited in Gikas & Grand 2013), SM 'tools in learning promotes a more student-centred course'. SM allows learners to network and work in partnership with each other and instructors.

Barczyk and Duncan (2011) identified elements that shape how Web 2.0 tools can be used meaningfully. Without structure, SM can negatively impact student learning. According to Barczyk and Duncan (2011), SM is used for educational purposes by lecturers at colleges and universities for learning and teaching. Studies by Barczyk and Duncan (2011) also indicated that more than half of the surveyed professors used tools such as videos, blogs, podcasts, video casts, and wikis in their classes. In his framework of SM, Kietzmann *et al.* (2011) identified seven building blocks for its functionality and usage: identity, conversation, sharing, presence, relationship, reputation and groups (Rosmalab & Rosmalab 2012). The diagram in Figure 1 illustrates the SM functionality framework.

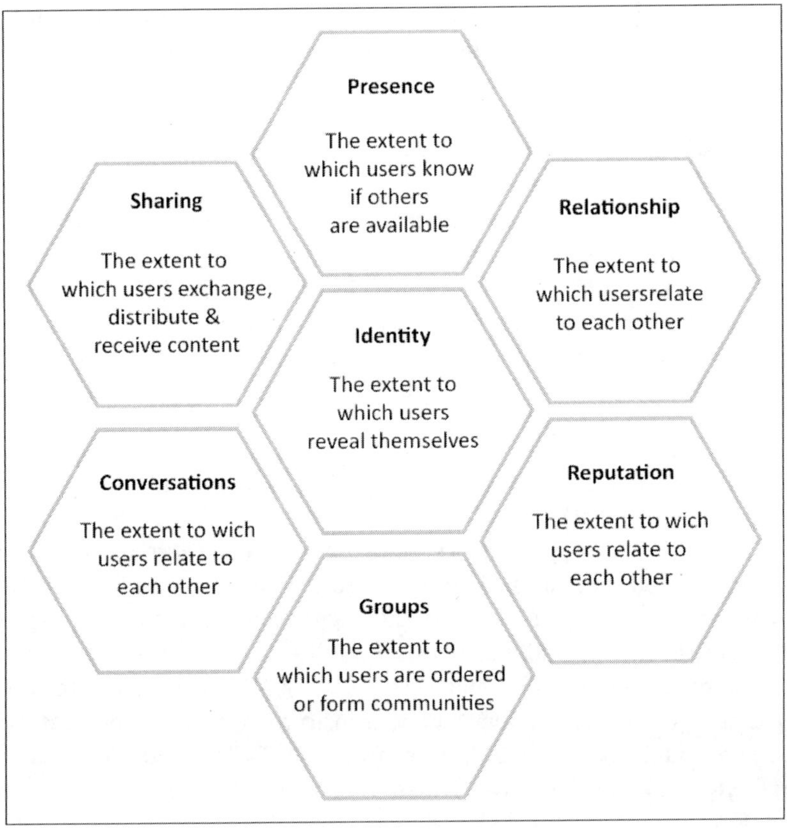

FIGURE 1: Social media functionality framework.

SM is not new in the learning and teaching sector, but it is still not widely accepted. It is important to note that if, and only if, the medium is used for educational purposes and following set regulations by students to supplement the existing medium, a positive result will be obtained within the FET sector (Dzvapatsva 2013). Dzvapatsva (2013) states that students with certain personality types, who might have proved to be difficult to engage, can benefit more from the use of electronic media.

For example, 'shy students' proved to have benefited more online than they had done through face-to-face interaction with the lecturer (Dzvapatsva 2013). However, the electronic medium should be used to supplement the already existing methods or mediums, and should not be a complete replacement. SM helps learners to collaborate and interact with each other, which are key aspects in higher education. Furthermore, SM facilitates informal learning (Ebner *et al.* 2010). Informal learning is described as contextualised, unorganised, but intentional (Gikas & Grant 2013). Mobile computing, which is an extension of electronic learning through mobile devices, can be used to do research, investigate, collect information and collaborate with other students through social networking sites (SNS). The information gathered will be used to supplement formal learning (Gikas & Grand 2013). Electronic platforms like SM and a KP allow students to access study material and communicate with classmates and instructors, no matter where they are (Gikas & Grand 2013). In addition, electronic platforms allow students to classify, control and evaluate existing knowledge, and successfully put together and communicate this as new knowledge.

Within the higher education sector, especially at FET colleges in South Africa, an information technology subject is found in most courses, either as Life Orientation (LO) or as Office Data Processing (ODP). These subjects focus on familiarising students with technology. Furthermore, they developing the academic skills required to locate specialised information, as well as for managing data, conducting online research, creating and managing documents and internet usage software. Lecturers in the field of Information Communication and Technology (ICT) should be aware of the fact that electronic technologies have transformed learning styles as most students would been using SM, for example FB, Twitter and Whatsapp, for a few years already, mainly by means of electronic devices, and lately, mobile devices.

Social media challenges within the FET sector

E-learning platforms like SM and KPs offer many advantages when compared to the traditional chalk-and-board methods. Some of the possible advantages, as noted by Cantoni, Cellario and Porta (2004), include:

- less expensive to deliver and faster
- provides consistent content, unlike in traditional methods where different lecturers can teach the same content differently
- it works from anywhere and anytime (virtual classroom)
- it can improve retention through varying types of content through videos and interaction

- it provides immediate feedback, especially if mobile devices are being used
- students are not afraid to make mistakes online so they can take risks and that alone can have a lasting positive effect (n.p.).

E-learning platforms are different from classroom-based learning; hence they require accurate development, monitoring and management. Electronic platforms, such as portals, require real subject experts to generate new skills in content and to verify the existing knowledge. It requires lecturers to be cognitively sensitive and well-sequenced. It also requires the explicit selection of instructional design to make learning experiences effective and long-lasting (Cantoni *et al.* 2004). Cantoni *et al.* (2004:336) state that electronic platforms may be 'intimidating, confusing or simply frustrating, lacking part of the informal social interaction and face-to-face contact of traditional classroom training'. This has been pointed out by some college lecturers who refused to use SM, citing them as disruptive. Another major concern is privacy issues, when platforms like FB are considered, which store an individual's details forever.

Because electronic learning platforms are free and can be used at liberty, they require self-discipline and more responsibility from students, which might be a challenge considering the type of students enrolled at FET colleges. Most of the students have social, economic and behavioural difficulties (SEBD).

Knowledge portals in higher education

Portal applications offer institutions a planned means of increasing their competitive advantage by encouraging innovation and research activities, which might lead to a greater success rate in grant acquisition, as well as growth in terms of prestige (Rajalakshmi & WahidiBanu 2009). According to Nonaka and Takeuchi (1995), knowledge creation comprises the utilisation of existing knowledge that is new to a group (tacit) or to an individual. The process becomes comprehensive when the collected knowledge is stored and can be retrieved in future for use by a group or an individual. A KP in this case offers the best option in capturing, sharing and transferring the knowledge; the knowledge acquired is of limited value if not captured, stored and shared. The ultimate goal for the organisation will be knowledge retention, which a KP offers.

RESEARCH METHODOLOGY

The research unfolded as a dual process. Action research was used in both phases. The first phase investigated how the use of FB could improve the quality and quantity of passes for core subjects. Two NC(V) classes with 45 students doing Computer Programming (CP) as a subject were used for the research. FB was used to contact the students doing CP outside normal teaching time. Lecturers created two FB groups and invited students randomly to use the platform. The one group used FB with lecturer facilitation and they formed the control group (Group A), whilst the other group (Group B) used FB without lecturer facilitation.

The second phase of this study focused on improving lecturer efficiency through the quality of assessments and the study or lecturer notes prepared by former and current lecturers, approved by the programme manager based on the assessment guideline document. Before the implementation of SM and the KP, students struggled during the final examinations owing to the lack of exposure to subject experts outside the college. This phase aimed to assess whether or not the application of the KP would assist in improving the performance of students. The sample assessments which lecturers used from the KP would assist in boosting the students' year mark and prepare them for the final summative examination. The KP was made public so that all lecturers could use it. They only needed to create an account and register on the local intranet server which hosted the portal. Further, a check on academic efficiency (if any) needed to be tabled and explanations given on whether the pass rates were in relation to portal usage or non-usage.

The subject moderators, in most cases post level two lecturers, would score each lecturer for moderation of the papers. A high score is regarded as full compliance to quality, which is necessary for the students to prepare for their summative tests. The KP served as a central location or repository of resource materials needed by each subject lecturer in respective subject areas. The moderation result, classified according to Department of Education grading, is shown in Table 1.

At the end of the year, the results of the lecturers who used the KP portal were compared to those who did not use the portal; thus, those of 2011 and preceding years. The intent was to use these scores as a baseline to measure variation (either positive or negative) after the implementation of the KP.

TABLE 1: Department of Education moderation gradings.

Category	Scores (%)
Strongly valid	90–100
Valid	80–89
Acceptable	70–79
Average	60–69
Sub-standard	Below 60

The portal prototype was adapted and extended to cater for lecturer interaction whilst on campus. Lecturer interaction was however limited to the campus, although they could access the portal offline. Availability of the KP offline meant that lecturers could familiarise themselves with the set of questions that the students could expect at the end of the year and the expected answers. Also, lecturers could streamline the study material to ensure that the material offered a good match to the national examinations at the end of the year. The KP would be updated on a regular basis to ensure that new trends in the IT sector were addressed. Rau, Gao and Wu (2008) found that students were motivated and enthusiastic to learn once they managed to master assessments set on the KP.

IMPLEMENTATION

The researchers intended to collect data from 45 students, but seven students dropped out of college along the way. Eventually, 38 questionnaire responses were collected. Data was also collected from the seven lecturers on their use of the KP. Descriptive statistics was used in the form of frequencies, averages and percentages to illustrate the results from the collected data.

Data collection for the research was twofold:

1. Responses from questionnaires distributed to students using SM: Two NC(V) level three classes were used for this study. The combined classes were given a baseline test before the introduction of SM. At a later stage, students were randomly split into two separate (virtual) classes for the purpose of this study. One group (Group A) used SM with lecturer facilitation; the other group (Group B) did not have lecturer facilitation. This random split was done to avoid possible bias. A FB group was created for use by Group A, whilst Group B studied with use of SM but without lecturer monitoring. This study aimed to determine how the increase in contact time through SM would assist the students in improving their performance. At the end of the term, a questionnaire was distributed to the remaining 38 students, and an analysis was done. Also, test scores for the subject were compared for the two groups.

2. Responses from lecturers and those students who used the KP: The KP used for the research was designed using open source software (PHP, MySQL and CSS) and hosted on the college's internal web server after seeking permission from the campus manager/principal. Lecturers were able to access it from any computer laboratory at the college and in their offices. All lecturers from the IT Department were asked to familiarise themselves with the portal and the information collected which was used for their assessment moderation. The results obtained from the moderation exercise were compared to those before the use of the KP.

FINDINGS

Increase in contact time

From Group A (with FB use), 83% (*n* = 15) indicated that FB helped to increase their contact time. Group B had six respondents who shared the same sentiments, which is 67.7% of the total number of respondents (the combined group). From the two groups it could be observed that the students generally agreed that FB could be useful for increasing contact time with the learning material, their peers, and the lecturers or subject experts.

Role of FB in learning computer programming

In Group A, 94.4% (A = 11; SA = 6; totalling 17 students) of the students stated that SM could be used to study CP. In Group B, 23% (*n* = 3) of the students were of the opinion that FB could be used to study CP, whilst 69.2% (*n* = 9) of the students from this group were neutral on SM usage in this regard. From these findings it is clear that more students, even those who did not use SM, agreed on the potential of SM to increase the contact time needed for practical subjects like CP. Ten students (55.6%) indicated that SM had positively changed their attitude towards studying CP and seven (38.9%) stated that FB increased their involvement towards studying the subject. Lesser percentages were obtained from Group B, namely 23.1%

($n = 4$) and 15.4% ($n = 3$) respectively. More importantly, the number of students who were satisfied by the responses they received on the FB platform, 67% ($n = 12$), was significant when compared to the 15.4% ($n = 2$) from Group B, with 53.8% offering a neutral response, and 30.8% ($n = 4$) stating that the responses from their fellow students were not satisfactory, nor reliable.

It is important to note here that the research was conducted in two phases. The one phase attempted to determine whether or not the use of a SM application would improve the contact time necessary for increasing performance, whilst the second phase concentrated on the quality of assessments within the NC(V) component of the Department of ICT at the FET college. All seven lecturers teaching ICT subjects used the KP for their assessments. Assessment quality is currently reflected 'as a moderation percentage obtained after the subject moderator has scrutinised both lecturer and student portfolios of evidence' (Dietrich *et al.* 2011). Table 2 represents the moderation scores supplied by the respondents for each of the listed subjects before the application of the KP.

TABLE 2: The moderation scores before the knowledge portal introduction.

Category	Scores (%)
Introduction to Information Systems	70–79
Introduction to Systems Development	70–79
Electronics	60–69
Systems Analysis L3	60–69
Hardware and Software	80–89
Principles of Computer Programming	70–79
Systems Analysis L4	70–79
Computer Programming	70–79
Data Communication and Networking	80–89

The KP prototype was launched after the first term moderation. All staff members were introduced to the KP and a training session was conducted during a staff development meeting. All lecturers were encouraged to comment freely through email as they interacted with the portal. For the lecturers, the portal included samples of lesson plans and explained how each lesson outcome is achieved when planning. Typical formative assessments included on the KP could be used by both lecturers and students to assess the content. The portal also provided a central store from which lecturers could access important subject content.

TABLE 3: The moderation scores after the knowledge portal introduction.

Moderation scores	0%–59%	60%–69%	70%–79%	80%–89%	90%–100%
Introduction to Information Systems	-	-	-	x	-
Introduction to Systems Development	-	-	-	-	x
Electronics	-	-	x	-	-
Systems Analysis level 3	-	-	-	x	-
Hardware and Software	-	-	-	x	-
Principles of Computer Programming	-	-	-	x	-
Systems Analysis level 4	-	-	-	x	-
Computer Programming	-	-	-	x	-
Data Communication and Networking	-	-	-	-	x

TABLE 4: The comparison of the moderation scores before and after the knowledge portal introduction.

Moderation scores	Before knowledge portal introduction					After knowledge portal introduction				
	0%–59%	60%–69%	70%–79%	80%–89%	90%–100%	0%–59%	60%–69%	70%–79%	80%–89%	90%–100%
Introduction to Information Systems	-	-	x	-	-	-	-	-	x	-
Introduction to Systems Development	-	-	x	-	-	-	-	-	-	x
Electronics	-	x	-	-	-	-	-	x	-	-
Systems Analysis level 3	-	x	-	-	-	-	-	-	x	-
Hardware and Software	-	-	-	x	-	-	-	-	x	-
Principles of Computer Programming	-	-	x	-	-	-	-	-	x	-
Systems Analysis level 4	-	-	x	-	-	-	-	-	x	-
Computer Programming	-	-	x	-	-	-	-	-	x	-
Data Communication and Networking	-	-	-	x	-	-	-	-	-	x

Table 3 illustrates average moderation scores for each of the subjects after three rounds of moderation.

It is clear from the results of the survey that the participants of the study had a positive experience with the KP prototype. Table 4 further highlights the positive variation of moderation scores achieved after the introduction of the portal.

As the results show, almost all the subjects, except Hardware and Software, showed an improvement in moderation results. However, it must be noted that even though Electronics showed an improvement of +/-10%, this is still regarded as below expected or outside the Valid Assessment range. The expected/valid assessment should score an average of between 80% and 89% during moderation. Introduction to Systems Development and Data Communication achieved average scores of above 90%.

Apart from initial technical difficulties experienced during the first phase roll-out of the portal, there were few challenges of any significance. This is partly due to the fact that interaction was at first restricted mainly to lecturers to determine whether or not it would affect the validity of their assessment instruments. However, when interaction was extended to include students, some problems (mainly involving IT infrastructure) occurred that were not as easy to avoid. Bandwidth was the biggest challenge, which the IT support technicians said was as a result of abuse by some students. There was also an increase in network activity as most students realised the importance of the KP. The other challenge which had to be addressed was access everywhere on the campus, apart from the computer laboratories. Despite these factors, there was a remarkable improvement in the throughput rate from six students graduating in 2011 to 18 students in 2012.

CONCLUSION

The research confirms and expands on previous research (Dietrich *et al.* 2011; Koles & Nagy 2012; Dzvapatsva 2013; Rau*et al.* 2008) on the potential of electronic platforms (SM and KP) to increase the performance of FET learners in South Africa. As stated by Gikas and Grand (2013), educators need to make use of emerging technology to teach the so-called Net Generation of students. In fact, Rosmalab and Rosmalab (2012) show that using diverse SM tools together to perform various tasks is more likely to increase individual knowledge acquisition. Learning through electronic platforms like SM and a KP indicates that valuable learning relies not only on social relations and association, but also on personal expression and information aggregation. Having seen the potential of SM and a KP, it should be emphasised that policies on these platforms need to be followed to obtain the best results from students, as they can be a problem if not well monitored (Cantoni *et al.* 2004:336).

This article assists in providing a glimpse into the countless ways in which electronic platforms might be used to provide the support needed for collaboration, networking and information sharing within FET institutions. Furthermore, this research will contribute towards a better understanding of how electronic platforms (i.e. SM and a KP) can be used in FET colleges in South Africa. This research confirms that the created KP was valuable not only as a knowledge repository, but also to support other processes. These processes include administrative issues and team building amongst staff members. The researchers hope that future research on the topic will build on, extend, test and complement the results which have been presented in this article.

LIMITATIONS OF THE STUDY

The research only focused on FB, without comparing it to other mediums. However, the data collected provides a good platform for future research as it shows the current trends in learning and teaching.

Final thoughts

Consistent with social media trends in general, ICT educators have a responsibility towards themselves (i.e. continuing professional development, joint research, academic or scholarly communication, publishing and producing peer reviewed articles), their students (improving their pass rates and empowering them to have more control over their own education), and the general public. Although SM and other electronic learning platforms offer opportunities, there is an increased need for training on how to use these technologies to improve and augment teaching and learning to support FET students and lecturers, with the ultimate objective of improving pass rates. However, if there are no clear policies on usage, the platforms can be abused by students.

COMPETING INTERESTS

The authors declare that they have no financial or personal relationship(s) that may have inappropriately influenced them in writing this article.

AUTHORS' CONTRIBUTIONS

G.P.D. (University of the Western Cape) and A.J.D. (University of the Western Cape) conducted the research and drafted the article, whilst Z.M. (University of the Western Cape) mentored and contributed significantly to the finalisation of the article.

REFERENCES

1. Bock, G.W. & Kim, Y.G., 2002, 'Breaking the myths of rewards: An exploratory study of attitudes about knowledge sharing', *Information Resource Management Journal* 15(2), 14–21.

2. Barczyk, C.C. & Duncan, D.G., 2011, 'Social networking media as a tool for teaching business administration courses',*International Journal of Humanities and Social Science* 1(17).

3. Cantoni, V., Cellario, M. & Porta, M., 2004, 'Perspectives and challenges in e-learning: Towards natural interaction paradigms', *Journal of Visual Languages and Computing* 15, 333–345.

4. Daft, R.L. & Lengel, R.H., 1986, 'Organizational information requirements, media richness and structural design', *Management Science* 32(5), 554–571.

5. Dzvapatsva, G.P., 2013, 'The use of social media as a means of improving the quantity and quality of the pass rate in computer programming at FET colleges in the Western Cape', Masters dissertation, University of the Western Cape.

6. Dietrich, A., Whyte, G. & Mitrovic, Z., 2011, 'E-learning at FET colleges: The application of a knowledge portal within the ICT Department of the College of Cape Town', *The ResNes Conference 2011*, East London, South Africa.

7. Ebner, M., Lienhardt, C., Rohs, M. & Meyer, I., 2010, 'Microblogs in Higher Education – A chance to facilitate informal and process-oriented learning?', *Computers & Education* 55(1), 92–100.

8. Hamid, S., Chang, S. & Kurnia, S., 2009, 'Identifying the use of online social networking in higher education', *ASCILITE 2009 conference*, December 2009, pp. 6–9.

9. Kabilan, K.M., Ahmad, N. & Abidin, M.J.Z., 2010, 'Facebook: An online environment for learning of English in institutions of higher education?' *Internet and Higher Education* 13, 179–187.

10. Kaplan, A.M. & Haenlein, M., 2010, 'Users of the world, unite! The challenges and opportunities of social media', *Business Horizons* 53(1), 59–68.

11. Kietzmann, J.H., Hermkens, K., McCarthy, I.P. & Silvestre, B.S., 2011, 'Social media? Get serious! Understanding the functional building blocks of social media', *Business Horizons* 54, 241–251.

12. Koles, B. & Nagy, P., 2012, 'Facebook usage patterns and school attitudes', *Multicultural Education & Technology Journal*6(1), 4-17.

13. Gikas, J. & Grant, M.M., 2013, 'Mobile computing devices in higher education: Student perspectives on learning with cellphones, smart phones and social media', *Internet and Higher Education* 19, 18–26.

14. Leung, C.F., 2002, 'Promoting an active learning environment within a congested curriculum', *International Conference on Engineering Education Proceedings*, Manchester, United Kingdom.

15. Nonaka, I. & Takeuchi, H., 1995, *The knowledge-creating company*, Oxford University Press, New York.

16. Rajalakshmi, S. & WahidiBanu, D., 2009, 'Developing an education web portal for knowledge sharing and capturing',*International Journal of Engineering and Technology* 1(3).

17. Rau, P.P., Gao, Q. & Wu, L., 2008, 'Using mobile communication technology in high school education: Motivation, pressure, and learning performance', *Computers & Education* 50, 1–22.

18. Rosmalab, F. & Rosmalab, D., 2012, 'Study of social networking usage in higher education environment', *3rd International*

Conference on e-Learning ICEL 2011, 23–24 November 2011, Bandung, Indonesia, *Procedia - Social and Behavioral Sciences* 67, 156–166.

19. Sarachan, J. & Reinson, K.F., 2011, 'Public issues, private concerns: Social media and course management systems in higher education', *Cutting-edge Technologies in Higher Education* 1, 227–244.

20. Short, J.W., Williams, E.E. & Christie, B., 1976, *The social psychology of telecommunications*, John Wiley, London.

Index

B

Big data analytics, 246, 252

C

collaboration, 43, 57, 60, 61, 245, 261, 307

D

Discovery, 47, 268

E

Education Institutions, 121, 122

F

Framework, 58, 119, 121, 122, 124, 126, 129

FUNCTIONALITY, 34

I

information handling, 57, 59, 60, 61, 64

Information Security, 121, 122, 130, 131, 132, 141, 142

K

knowledge economy, 57, 235

P

Predictive analytics, 246, 262

R

research, 2, 3, 5, 7, 8, 18, 20

U

Unstructured data analytics, 246